Fundamentals of Data Warehouses

Matthias Jarke • Maurizio Lenzerini
Yannis Vassiliou • Panos Vassiliadis

Fundamentals of Data Warehouses

Second, Revised and Extended Edition

With 59 Figures

Springer

Matthias Jarke
Dept. of Computer Science V
RWTH Aachen
Ahornstraße 55
52056 Aachen, Germany
jarke@informatik.rwth-aachen.de

Maurizio Lenzerini
Dipartimento di Informatica
e Sistemistica
Università di Roma "La Sapienza"
Via Saleria 113
00198 Rome, Italy
lenzerini@dis.uniromal.it

Yannis Vassiliou
Panos Vassiliadis
Dept. of Electrical and Computer
Engineering
Computer Science Division
National Technical University of Athens
15773 Zographou
Athens, Greece
yv@cs.ntua.gr

ACM Computing Classification (1998): H.2.7, H.2-4, D.2.9, I.2, K.6.3-4

Library of Congress Cataloging-in-Publication Data

Fundamentals of data warehouses / Matthias jarke ... [et al.]
p. cm.
Includes bibliographical references and index.
ISBN 354020894
1. Data warehousing. I Jarke, Matthias.
QA76.9.D37 F86 2002
658.4'038'0285574—dc21

2002070677

ISBN 3-540-42089-4 Springer-Verlag Berlin Heidelberg New York
ISBN 3-540-65365-1 1 st ed. Springer-Verlag Berlin Heidelberg New York

Cover design: KünkelLopka, heidelberg

**Jarke et. al.: Fundamentals of Data Warehouses,
Second, Revised & Extended edition**

First Indian Reprint 2008
Second Indian Reprint 2009

ISBN 978-81-8128-914-8

This edition is manufactured in India for sale only in India, Pakistan, Bangladesh, Nepal and Sri Lanka and any other country as authorized by the publisher.

This edition is published by Springer (India) Private Limited, (a part of Springer Science+Business Media, Registered Office: 3rd Floor, Gandharva Mahavidyalaya, 212, Deen Dayal Upadhyaya Marg, New Delhi – 110 002, India.

Printed in India by Rashtriya Printers, Delhi.

Preface to the Second Edition

Data warehousing is still an active and rapidly evolving field even though many of the foundations are stabilizing. The first edition of this book sold out within little more than a year. Moreover, we found that a number of updates had to be made in particular to the state of the practice because some of the tools described in the first edition are no longer on the market and a trend towards more integrated solutions can be observed. We are grateful to several researchers and developers from data warehouse vendor companies who pointed out such issues. In addition, the second edition contains more information about new developments in metadata management and, in a new Chap. 8, a comprehensive description and illustration of the quality-oriented data warehouse design and operation methodology developed in the final stages of the European DWQ project. Thanks to Ingeborg Mayer of Springer-Verlag for her persistence and support in the revision for this edition. Many Thanks to Ulrike Drechsler and Christian Seeling for their technical support of this revision.

Aachen,	Matthias Jarke,
Athens,	Maurizio Lenzerini,
Rome,	Yannis Vassiliou,
September 2002	Panos Vassiliadis

Preface to the First Edition

This book is an introduction and sourcebook for practitioners, graduate students, and researchers interested in the state of the art and the state of the practice in data warehousing. It resulted from our observation that, while there are a few hands-on practitioner books on data warehousing, the research literature tends to be fragmented and poorly linked to the commercial state of practice. As a result of the synergistic view taken in the book, the last chapter presents a new approach for data warehouse quality assessment and quality-driven design which reduces some of the recognized shortcomings. For the reader, it will be useful to be familiar with the basics of the relational model of databases to be able to follow this book.

The book is made up of seven chapters. Chapter 1 sets the stage by giving a broad overview of some important terminology and vendor strategies. Chapter 2 summarizes the research efforts in data warehousing and gives a short description of the framework for data warehouses used in this book.

The next two chapters address the main data *integration issues* encountered in data warehousing. Chapter 3 presents a survey of the main techniques used when linking information sources to a data warehouse, emphasizing the need for semantic modeling of the relationships. Chapter 4 investigates the propagation of updates from operational sources through the data warehouse to the client analyst, looking both at incremental update computations and at the many facets of refreshment policies.

The next two chapters study the client-side of a data warehouse. Chapter 5 shows how to reorganize relational data into the multidimensional data models used for online analytic processing applications, focusing on the conceptualization of, and reasoning about multiple, hierarchically organized dimensions. Chapter 6 takes a look at query processing and its optimization, taking into account the reuse of materialized views and the multidimensional storage of data.

In the literature, there is not much coherence among all these technical issues on the one side, and the business reasoning and design strategies underlying data warehousing projects. Chapter 7 ties these aspects together. It presents an extended architecture for data warehousing and links it to explicit models of data warehouse quality. It is shown how this extended approach can be used to document the quality of a data warehouse project and to design a data warehouse solution for specific quality criteria.

The book resulted from the ESPRIT Long Term Research Project DWQ (Foundations of Data Warehouse Quality) which was supported by the Commission of the European Union from 1996 to 1999. DWQ's goal was to develop a semantic foundation that will allow the designers of data warehouses to link the choice of deeper models, richer data structures, and rigorous implementation techniques to quality-of-service factors in a systematic manner, thus improving the design, the operation, and most importantly the long-term evolution of data warehouse applications.

Many researchers from all DWQ partner institutions – the National Technical University of Athens (Greece), RWTH Aachen University of Technology (Germany), DFKI German Research Center for Artificial Intelligence, the INRIA National Research Center (France), IRST Research Center in Bolzano (Italy), and the University of Rome – La Sapienza, have contributed to the underlying survey work. Their contributions are listed in the following overview. Great thanks go to our industrial collaborators who provided product information and case studies, including but not limited to Software AG, the City of Cologne, Team4 Systemhaus, Swiss Life, Telecom Italia, and Oracle Greece. Valuable comments from our EU project officer, David Cornwell, as well as from the project reviewers Stefano Ceri, Laurent Vieille, and Jari Veijalainen have sharpened the presentation of this material. Last but not least we thank Dr. Hans Wössner and his team at Springer-Verlag for a smooth production process. Christoph Quix was instrumental in supporting many of the technical editing tasks for this book; special thanks go to him.

Aachen,	Matthias Jarke,
Athens,	Maurizio Lenzerini,
Rome,	Yannis Vassiliou,
June 1999	Panos Vassiliadis

Overview and Contributors

Addresses of Contributors

Matthias Jarke, Christoph Quix
Informatik V (Information Systems), RWTH Aachen
Ahornstr. 55, 52056 Aachen, Germany
Email: jarke@informatik.rwth-aachen.de

Franz Baader, Ulrike Sattler
Institut für Theoretische Informatik, TU Dresden,
01062 Dresden, Germany
Email: baader@tcs.inf.tu-dresden.de

Yannis Vassiliou, Spyros Ligoudistianos, Timos Sellis, Aris Tsois, Panos Vassiliadis
Department of Electrical and Computer Engineering
Computer Science Division, National Technical University of Athens
Zographou 157 73, Athens, Greece
Email: yv@cs.ntua.gr

Enrico Franconi
University of Manchester, Department of Computer Science
Oxford Rd., Manchester M13 9PL, United Kingdom
Email: franconi@cs.man.ac.uk

Eric Simon, Mokrane Bouzeghoub, Françoise Fabret, Helena Galhardas,
Maja Matulovic-Broqué, Jaoa Pereira
Project Rodin, INRIA Rocquencourt
Domaine de Voluceau BP.105, 78153 Le Chesnay Cedex, France
Email: Eric.Simon@inria.fr

Maurizio Lenzerini, Diego Calvanese, Giuseppe De Giacomo, Daniele Nardi,
Ricardo Rosati
Dipartimento di Informatica e Sistemistica, Università di Roma "La Sapienza"
Via Salaria 113, 00198 Roma, Italy
Email: lenzerini@dis.uniroma1.it

Werner Nutt
Department of Computing and Electrical Engineering, Heriot-Watt University
Riccarton, Edinburgh EH14 4AS, United Kingdom
Email: nutt@cee.hw.ac.uk

Manfred A. Jeusfeld
KUB Tilburg, INFOLAB
Warandelaan 2, Postbus 90153, 5000 LE Tilburg, The Netherlands
Email: jeusfeld@kub.nl

Contents

1 Data Warehouse Practice: An Overview

Since the beginning of data warehousing in the early 1990s, an informal consensus has been reached concerning the major terms and components involved in data warehousing. In this chapter, we first explain the main terms and components. Data warehouse vendors are pursuing different strategies in supporting this basic framework. We review a few of the major product families and the basic problem areas data warehouse practice and research are faced with today.

A data warehouse (DW) is a collection of technologies aimed at enabling the knowledge worker (executive, manager, and analyst) to make better and faster decisions. It is expected to have the right information in the right place at the right time with the right cost in order to support the right decision. Traditional online transaction processing (OLTP) systems are inappropriate for decision support and high-speed networks cannot, by themselves, solve the information accessibility problem. Data warehousing has become an important strategy to integrate heterogeneous data sources and to enable online analytic processing (OLAP).

A report from the META Group in 1996 predicted data warehousing would be a US $ 13 000 million industry within two years ($8000 million on hardware, $ 5000 million on services and systems integration), while 1995 represented $ 2000 million in expenditures. In 1998, reality had exceeded these figures, reaching sales of $14 600 million. By 2000, the subsector of OLAP alone exceeded $ 2500 million. Table 1.1 differentiates the trends by product sector.

Table 1.1. Estimated sales in millions of dollars [ShTy98] (* Estimates are from [PeCr00])

	1998	1999	2000	2001	2002	CAGR (%)
RDBMS sales for DW	900.0	1110.0	1390.0	1750.0	2200.0	25.0
Data Marts	92.4	125.0	172.0	243.0	355.0	40.0
ETL tools	101.0	125.0	150.0	180.0	210.0	20.1
Data Quality	48.0	55.0	64.5	76.0	90.0	17.0
Metadata Management	35.0	40.0	46.0	53.0	60.0	14.4
OLAP (including implementation services)*	2000	2500	3000	3600	4000	18.9

The number and complexity of projects – with project sizes ranging from a few hundred thousand to multiple millions of dollars – is indicative of the difficulty of designing good data warehouses. Their expected duration highlights the need for documented quality goals and change management. The emergence of data warehousing was initially a consequence of the observation by W. Inmon and E. F. Codd in the early 1990s that operational-level online transaction processing (OLTP) and decision support applications (OLAP) cannot coexist efficiently in the same database environment, mostly due to their very different transaction characteristics. Meanwhile, data warehousing has taken a much broader role, especially in the context of reengineering legacy systems or at least saving legacy data. Here, DWs are seen as a strategy to bring heterogeneous data together under a common conceptual and technical umbrella and to make them available for new operational or decision support applications.

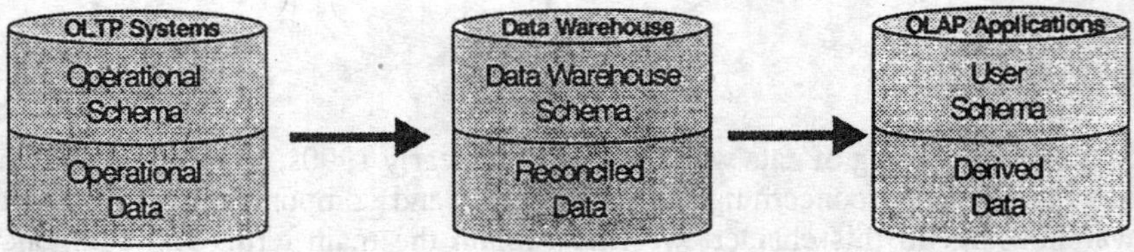

Fig. 1.1. Data warehouses: a buffer between transaction processing and analytic processing

A data warehouse caches selected data of interest to a customer group, so that access becomes faster, cheaper, and more effective (Fig. 1.1). As the long-term buffer between OLTP and OLAP, data warehouses face two essential questions: how to reconcile the stream of incoming data from multiple heterogeneous legacy sources, and how to customize the derived data storage to specific OLAP applications. The trade-off driving the design decisions concerning these two issues changes continuously with business needs. Therefore, design support and change management are of greatest importance if we do not want to run DW projects into dead ends.

Vendors agree that data warehouses cannot be off-the-shelf products but must be designed and optimized with great attention to the customer situation. Traditional database design techniques do not apply since they cannot deal with DW-specific issues such as data source selection, temporal and aggregated data, and controlled redundancy management. Since the wide variety of product and vendor strategies prevents a low-level solution to these design problems at acceptable costs, serious research and development efforts continue to be necessary.

1.1 Data Warehouse Components

Figure 1.2 gives a rough overview of the usual data warehouse components and their relationships. Many researchers and practitioners share the understanding that a data warehouse architecture can be understood as layers of materialized views on top of each other. Since the research problems are largely formulated from this perspective, we begin with a brief summary description.

A data warehouse architecture exhibits various layers of data in which data from one layer are derived from data of the lower layer. *Data sources*, also called *operational databases*, form the lowest layer. They may consist of structured data stored in open database systems and legacy systems or unstructured or semi-structured data stored in files. The data sources can be either part of the operational environment of an organization or external, produced by a third party. They are usually heterogeneous, which means that the same data can be represented differently, for instance through different database schemata, in the sources.

Fig. 1.2. A generic data warehouse architecture

The central layer of the architecture is the *"global" data warehouse*, sometimes called *primary* or *corporate data warehouse*. According to Inmon [Inmo96], it is a "collection of integrated, nonvolatile, subject-oriented databases designed to support the decision support system (DSS) function, where each unit of data is relevant to some moment in time, it contains atomic data and lightly summarized data." The global data warehouse keeps a historical record of data. Each time it is changed, a new integrated snapshot of the underlying data sources from which it is derived is placed in line with the previous snapshots. Typically, the data warehouse may contain data that can be many years old (a frequently cited average age is two years). Researchers often assume (realistically) that the global warehouse consists of a set of materialized relational views. These views are defined in terms of other relations that are themselves constructed from the data stored in the sources.

The next layer of views are the "local" warehouses, which contain highly aggregated data derived from the global warehouse, directly intended to support activities such as informational processing, management decisions, long-term decisions, historical analysis, trend analysis, or integrated analysis. There are various kinds of local warehouses, such as the *data marts* or the *OLAP databases*. Data marts are small data warehouses that contain only a subset of the enterprise-wide data warehouse. A data mart may be used only in a specific department and contains only the data which is relevant to this department. For example, a data mart for the marketing department should include only customer, sales, and product information whereas the enterprise-wide data warehouse could also contain information on employees, departments, etc. A data mart enables faster response to queries because the volume of the managed data is much smaller than in the data warehouse and the queries can be distributed between different machines. Data marts may use relational database systems or specific multidimensional data structures.

There are two major differences between the global warehouse and local data marts. First, the global warehouse results from a complex extraction-integration-transformation process. The local data marts, on the other hand, result from an extraction/aggregation process starting from the global warehouse. Second, data in the global warehouse are detailed, voluminous (since the data warehouse keeps data from previous periods of time), and lightly aggregated. On the contrary, data in the local data marts are highly aggregated and less voluminous. This distinction has a number of consequences both in research and in practice, as we shall see throughout the book.

In some cases, an intermediate layer, called an *operational data store* (ODS), is introduced between the operational data sources and the global data warehouse. An ODS contains subject-oriented, collectively integrated, volatile, current valued, and detailed data. The ODS usually contains records that result from the transformation, integration, and aggregation of detailed data found in the data sources, just as for a global data warehouse. Therefore, we can also consider that the ODS consists of a set of materialized relational views. The main differences with a data warehouse are the following. First, the ODS is subject to change much more frequently than a data warehouse. Second, the ODS only has fresh and current data. Finally, the aggregation in the ODS is of small granularity: for example, the data can be weakly summarized. The use of an ODS, according to Inmon [Inmo96], is justified for corporations that need collective, integrated operational data. The ODS is a good support for activities such as collective operational decisions, or immediate corporate information. This usually depends on the size of the corporation, the need for immediate corporate information, and the status of integration of the various legacy systems. Figure 1.2 summarizes the different layers of data.

All the data warehouse components, processes, and data are – or at least should be – tracked and administered from a metadata repository. The metadata repository serves as an aid both to the administrator and the designer of a data warehouse. Since the data warehouse is a very complex system, its architecture (physical components, schemata) can be complicated; the volume of data is vast; and the processes employed for the extraction, transformation, cleaning, storage, and aggregation of data are numerous, sensitive to changes, and vary in time.

1.2 Designing the Data Warehouse

The design of a data warehouse is a difficult task. There are several problems designers have to tackle. First of all, they have to come up with the semantic reconciliation of the information lying in the sources and the production of an enterprise model for the data warehouse. Then, a logical structure of relations in the core of data warehouse must be obtained, either serving as buffers for the refreshment process or as persistent data stores for querying or further propagation to data marts. This is not a simple task by itself; it becomes even more complicated since the physical design problem arises: the designer has to choose the physical tables, processes, indexes, and data partitions, representing the logical data warehouse schema and facilitating its functionality. Finally, hardware selection and software development is another process that has to be planned from the data warehouse designer [AdVe98, ISIA97, Simo98].

It is evident that the schemata of all the data stores involved in a data warehouse environment change rapidly: the changes of the business rules of a corporation affect both the source schemata (of the operational databases) and the user requirements (and the schemata of the data marts). Consequently, the design of a data warehouse is an ongoing process, which is performed iteratively throughout the lifecycle of the system [KRRT98].

There is quite a lot of discussion about the methodology for the design of a data warehouse. The two major methodologies are the top-down and the bottom-up approaches [Kimb96, KRRT98, Syba97]. In the top-down approach, a global enterprise model is constructed, which reconciles the semantic models of the sources (and later, their data). This approach is usually costly and time-consuming; nevertheless it provides a basis over which the schema of the data warehouse can evolve. The bottom-up approach focuses on the more rapid and less costly development of smaller, specialized data marts and their synthesis as the data warehouse evolves.

No matter which approach is followed, there seems to be agreement on the general idea concerning the final schema of a data warehouse. In a first layer, the ODS serves as an intermediate buffer for the most recent and detailed information from the sources. The data cleaning and transformation is performed at this level. Next, a database under a denormalized "star" schema usually serves as the central repository of data. A *star schema* is a special-purpose schema in data warehouses that is oriented towards query efficiency at the cost of schema normalization (cf. Chap. 5 for a detailed description). Finally, more aggregated views on top of this star schema can also be precalculated. The OLAP tools can communicate either with the upper levels of the data warehouse or with the customized data marts: we shall detail this issue in the following sections.

1.3 Getting Heterogeneous Data into the Warehouse

Data warehousing requires access to a broad range of *information sources*:

- Database systems (relational, object-oriented, network, hierarchical, etc.)
- External information sources (information gathered from other companies, results of surveys)
- Files of standard applications (e.g., Microsoft Excel, COBOL applications)
- Other documents (e.g., Microsoft Word, World Wide Web)

Wrappers, loaders, and mediators are programs that load data of the information sources into the data warehouse. Wrappers and loaders are responsible for loading, transforming, cleaning, and updating the data from the sources to the data warehouse. Mediators integrate the data into the warehouse by resolving inconsistencies and conflicts between different information sources. Furthermore, an extraction program can examine the source data to find reasons for conspicuous items, which may contain incorrect information [BaBM97].

These tools – in the commercial sector classified as Extract-Transform-Load (ETL) tools – try to automate or support tasks such as [Gree97]:

- Extraction (accessing different source databases)
- Cleaning (finding and resolving inconsistencies in the source data)
- Transformation (between different data formats, languages, etc.)
- Loading (loading the data into the data warehouse)
- Replication (replicating source databases into the data warehouse)
- Analyzing (e.g., detecting invalid/unexpected values)
- High-speed data transfer (important for very large data warehouses)
- Checking for data quality, (e.g., for correctness and completeness)
- Analyzing metadata (to support the design of a data warehouse)

1.4 Getting Multidimensional Data out of the Warehouse

Relational database management systems (RDBMS) are most flexible when they are used with a normalized data structure. Because normalized data structures are non-redundant, normalized relations are useful for the daily operational work. The database systems used for this role, so called OLTP systems, are optimized to support small transactions and queries using primary keys and specialized indexes.

While OLTP systems store only current information, data warehouses contain historical and summarized data. These data are used by managers to find trends and directions in markets, and supports them in decision making. OLAP is the technology that enables this exploitation of the information stored in the data warehouse.

Due to the complexity of the relationships between the involved entities, OLAP queries require multiple join and aggregation operations over normalized relations, thus overloading the normalized relational database.

Typical operations performed by OLAP clients include [ChDa97]:

- *Roll up* (increasing the level of aggregation)
- *Drill down* (decreasing the level of aggregation)
- *Slice* and *dice* (selection and projection)
- *Pivot* (reorienting the multidimensional view)

Beyond these basic OLAP operations, other possible client applications on data warehouses include:

- Report and query tools
- Geographic information systems (GIS)
- Data mining (finding patterns and trends in the data warehouse)
- Decision support systems (DSS)
- Executive information systems (EIS)
- Statistics

The OLAP applications provide users with a multidimensional view of the data, which is somewhat different from the typical relational approach; thus their operations need special, customized support. This support is given by multidimensional database systems and relational OLAP servers.

The database management system (DBMS) used for the data warehouse itself and/or for data marts must be a high-performance system, which fulfills the requirements for complex querying demanded by the clients. The following kinds of DBMS are used for data warehousing [Weld97]:

- Super-relational database systems
- Multidimensional database systems

Super-relational database systems. To make RDBMS more useful for OLAP applications, vendors have added new features to the traditional RDBMS. These so-called *super-relational* features include support for extensions to storage formats, relational operations, and specialized indexing schemes. To provide fast response time to OLAP applications, the data are organized in a *star* or *snowflake* schema (see also Chap. 5).

The resulting data model might be very complex and hard to understand for end users. Vendors of relational database systems try to hide this complexity behind special engines for OLAP. The resulting architecture is called *Relational OLAP* (ROLAP). In contrast to predictions in the mid-1990s, ROLAP architectures have not been able to capture a large share of the OLAP market. Within this segment, one of the leaders is MicroStrategy [MStr97] whose architecture is shown in Fig. 1.4. The RDBMS is accessed through *VLDB (very large databases) drivers*, which are optimized for large data warehouses.

The *DSS Architect* translates relational database schemas to an intuitive multidimensional model, so that users are shielded from the complexity of the relational data model. The mapping between the relational and the multidimensional data models is done by consulting the metadata. The system is controlled by the *DSS Administrator*. With this tool, system administrators can fine-tune the database schema, monitor the system performance, and schedule batch routines.

The *DSS Server* is a ROLAP server, based on a relational database system. It provides a multidimensional view of the underlying relational database. Other features are the ability to cache query results, the monitoring and scheduling of queries, and generating and maintaining dynamic relational data marts. *DSS Agent, DSS Objects,* and *DSS Web* are interfaces to end users, programming languages, or the World Wide Web.

Other ROLAP servers are offered by Red Brick [RBSI97] (subsequently acquired by Informix, then passed on to IBM) and Sybase [Syba97]. The Red Brick system is characterized by an industry-leading indexing and join technology for star schemas (Starjoin); it also includes a data mining option to find patterns, trends, and relationships in very large databases. They argue that data warehouses need to be constructed in an incremental, bottom-up fashion. Therefore, such vendors focus on support of distributed data warehouses and data marts.

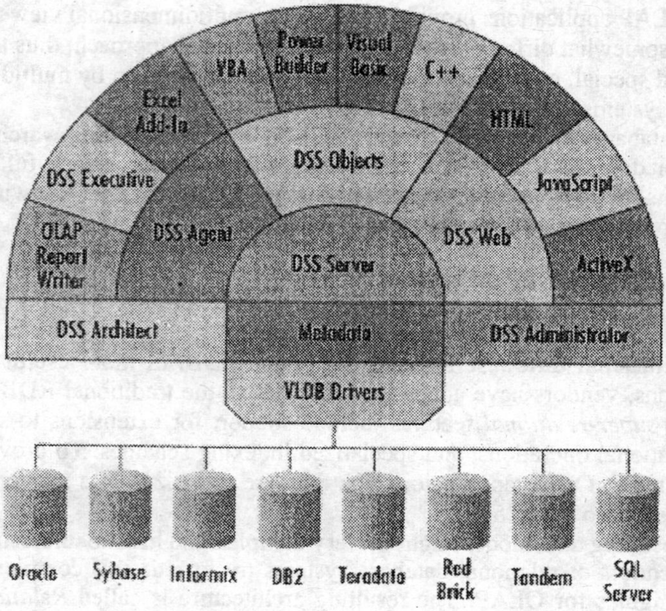

Fig. 1.3. MicroStrategy solution [MStr97]

Multidimensional database systems (MDDB) support directly the way in which OLAP users visualize and work with data. OLAP requires an analysis of large volumes of complex and interrelated data and viewing that data from various perspectives [Kena95]. MDDB store data in *n*-dimensional cubes. Each dimension represents a user perspective. For example, the sales data of a company may have the dimensions product, region, and time. Because of the way the data is stored, there are no join operations necessary to answer queries which retrieve sales data by one of these dimensions. Therefore, for OLAP applications, MDDB are often more efficient than traditional RDBMS [Coll96]. A problem with MDDB is that restructuring is much more expensive than in a relational database. Moreover, there is currently no standard data definition language and query language for the multidimensional data model.

In practical multidimensional OLAP products, two market segments can be observed [PeCr00]. At the low end, *desktop OLAP systems* such as Cognos PowerPlay, Business Objects, or Brio focus on the efficient and user-friendly handling of relatively small data cubes on client systems. Here, the MDBS is implemented as a data retailer [Sahi96]: it gets its data from a (relational) data warehouse and offers analysis functionality to end users. As shown in Fig. 1.5, ad-hoc queries are sent directly to the data warehouse, whereas OLAP applications work on the more appropriate, multidimensional data model of the MDDB. Market leaders in this segment support hundreds of thousands of workplaces.

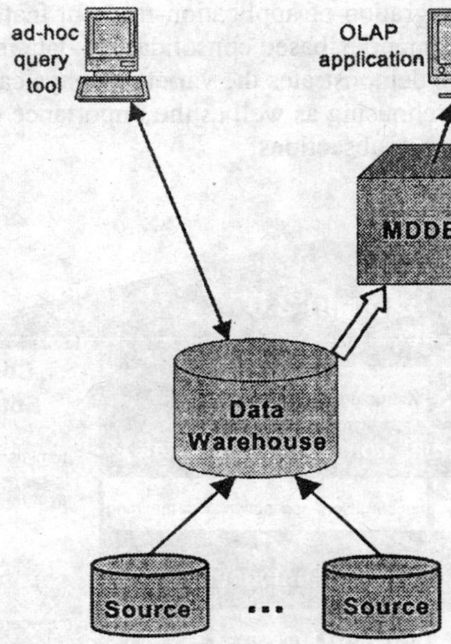

Fig. 1.4. MDDB in a data warehouse environment

At the high end, hybrid OLAP (HOLAP) solutions aim to provide full integration of relational data warehouse solutions (aiming at scalability) and multidimensional solutions (aiming at OLAP efficiency) in complex architectures. Market leaders include Hyperion Essbase, Oracle Express, and Microsoft OLAP.

Application-oriented OLAP. As pointed out by Pendse and Creeth [PeCr00], only a few vendors can survive on generic server tools as mentioned above. Many more market niches can be found for specific application domains. Systems in this sector often provide lots of application-specific functionality in addition to (or on top of) multidimensional OLAP (MOLAP) engines. Generally speaking, application domains can be subdivided into four business functions:

- Reporting and querying for standard controlling tasks
- Problem and opportunity analysis (often called Business Intelligence)
- Planning applications
- One-of-a-kind data mining campaigns or analysis projects

Two very important application domains are sales analysis and customer relationship management on the one hand, and budgeting, financial reporting, and consolidation on the other. Interestingly, only a few of the tools on the market are able to integrate the reporting and analysis stage for the available data, with the planning tasks for the future.

As an example, Fig. 1.6 shows the b2brain architecture by Thinking Networks AG [Thin01], a MOLAP-based environment for financial reporting and planning data warehouses. It shows some typical features of advanced application-oriented OLAP environments such as efficient custom-tailoring to new applications within a domain using metadata, linkage to heterogeneous sources and clients also via the

Internet, and seamless integration of application-relevant features such as heterogeneous data collection, semantics-based consolidation, data mining and planning. Therefore, the architecture demonstrates the variety of physical structures encountered in high-end data warehousing as well as the importance of metadata, both to be discussed in the following subsections.

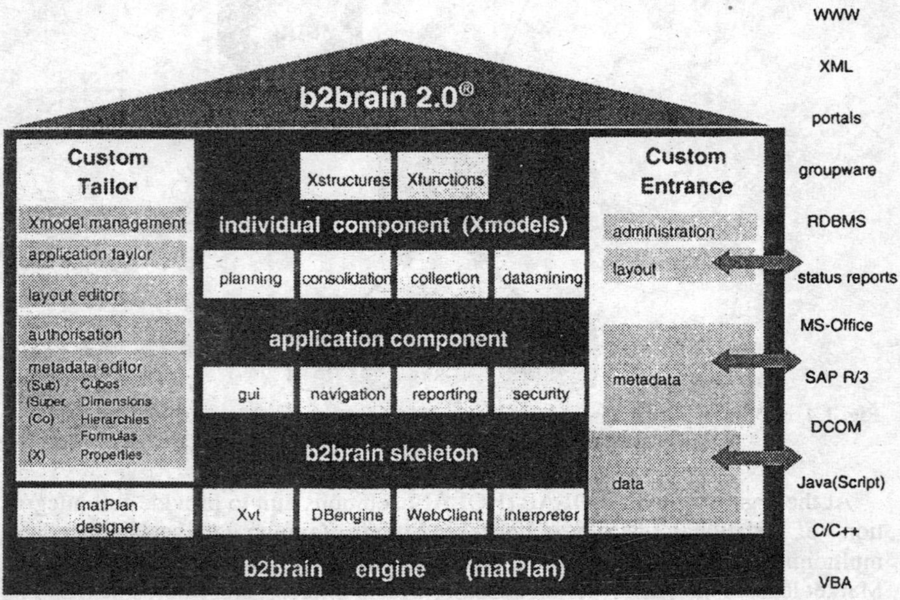

Fig. 1.5. Example of a DW environment for integrated financial reporting and planning

1.5 Physical Structure of Data Warehouses

There are three basic architectures for a data warehouse [Weld97, Muck96]:

- Centralized
- Federated
- Tiered

In a *centralized* architecture, there exists only one data warehouse which stores all data necessary for business analysis. As already shown in the previous section, the disadvantage is the loss of performance compared to distributed approaches. All queries and update operations must be processed in one database system.

On the other hand, access to data is uncomplicated because only one data model is relevant. Furthermore, building and maintaining a *central* data warehouse is easier than in a distributed environment. A central data warehouse is useful for companies, where the existing operational framework is also centralized (Fig. 1.7).

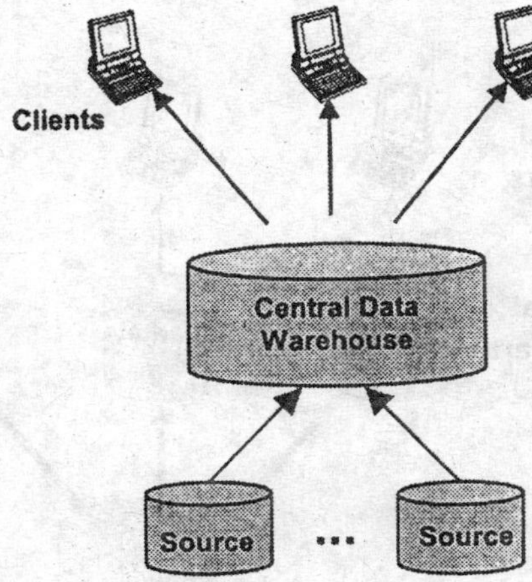

Fig. 1.6. Central architecture

A decentralized architecture is only advantageous if the operational environment is also distributed. In a *federated* architecture, the data is logically consolidated but stored in separate physical databases at the same or at different physical sites (Fig. 1.8). The local data marts store only the relevant information for a department. Because the amount of data is reduced in contrast to a central data warehouse, the local data mart may contain all levels of detail so that detailed information can also be delivered by the local system.

An important feature of the federated architecture is that the logical warehouse is only virtual. In contrast, in a *tiered* architecture (Fig. 1.9), the central data warehouse is also physical. In addition to this warehouse, there exist local data marts on different tiers which store copies or summaries of the previous tier but not detailed data as in a federate architecture.

There can be also different tiers at the source side. Imagine, for example, a super market company collecting data from its branches. This process cannot be done in one step because many sources have to be integrated into the warehouse. On the first level, the data of all branches in one region is collected, and in the second level, the data from the regions is integrated into one data warehouse.

The advantages of the distributed architecture are (a) faster response time because the data is located closer to the client applications and (b) reduced volume of data to be searched. Although, several machines must be used in a distributed architecture, this may result in lower hardware and software costs because not all data must be stored at one place and queries are executed on different machines. A scalable architecture is very important for data warehousing. Data warehouses are not static systems but evolve and grow over time. Because of this, the architecture chosen to build a data warehouse must be easy to extend and to restructure.

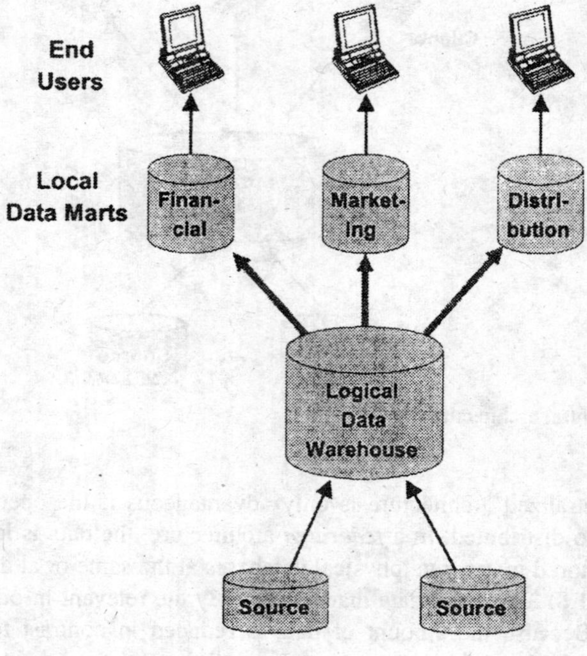

Fig. 1.7. Federated architecture

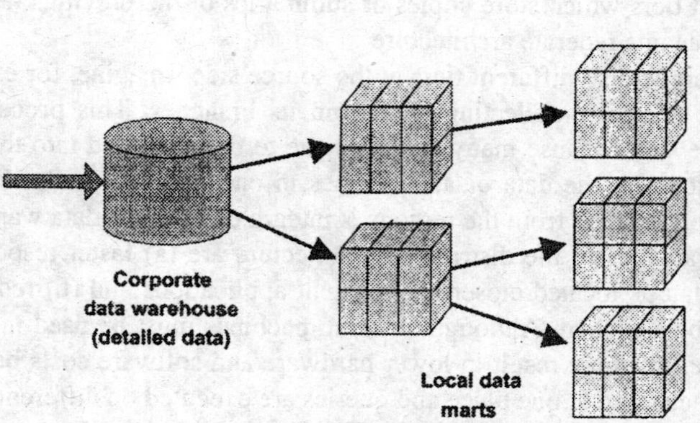

Fig. 1.8. Tiered architecture

1.6 Metadata Management

Metadata play an important role in data warehousing. Before a data warehouse can be accessed efficiently, it is necessary to understand what data is available in the warehouse and where is the data located In addition to locating the data that the end users require, metadata repositories may contain [AdCo97, MStr95, Micr96]:

- Data dictionary: contains definitions of the databases being maintained and the relationships between data elements
- Data flow: direction and frequency of data feed
- Data transformation: transformations required when data is moved
- Version control: changes to metadata are stored
- Data usage statistics: a profile of data in the warehouse
- Alias information: alias names for a field
- Security: who is allowed to access the data

As shown in Fig. 1.2, metadata is stored in a repository, where it can be accessed from every component of the data warehouse. Because metadata is used and provided by all components of the warehouse, a standard interchange format for metadata is necessary. The Metadata Coalition has proposed a *Metadata Interchange Specification* [MeCo96]; additional emphasis has been placed on this area through Microsoft's introduction of a repository product in their Office product suite, including some information models for data warehousing [BBC*99].

1.7 Data Warehouse Project Management

Projects concerning the design and operation of data warehouses tend to be of medium to very large size and may have significant impact on the organization. A number of authors have studied the cost drivers and risks of such projects. For an overview and some experiences with real-world projects, see [Vass00, PeCr00].

According to Shilakes and Tylman [ShTy98], the average time for the construction of a data warehouse is 12–36 months and the average cost for its implementation is between $ 1 million and $ 1.5 million. Data marts are a less risky expenditure, since they cost hundreds of thousands of dollars and take less than 12 months to implement. Inmon [Inmo97] offers some figures how these costs are spread over the different development and administration tasks (cf. Fig. 1.10).

In [Dema97], reasons for *failure in data warehousing projects* are discussed. As for *design factors*, there is an obvious deficit in textbook methodologies for data warehouse design. Proprietary solutions from vendors or do-it-yourself advice from experts seem to define the landscape. The *technical factors* reveal a gap in the evaluation and choice of hardware components. As one can see in Fig. 1.10, hardware costs up to 60% of a data warehouse budget (disk, processor and network costs). Critical software (DBMS and client tools) which is purchased (and not developed in-house) takes up to 16% of the budget. The fact that the average size of data warehouses increases year by year makes the problem even tougher, as does the increasing number of users.

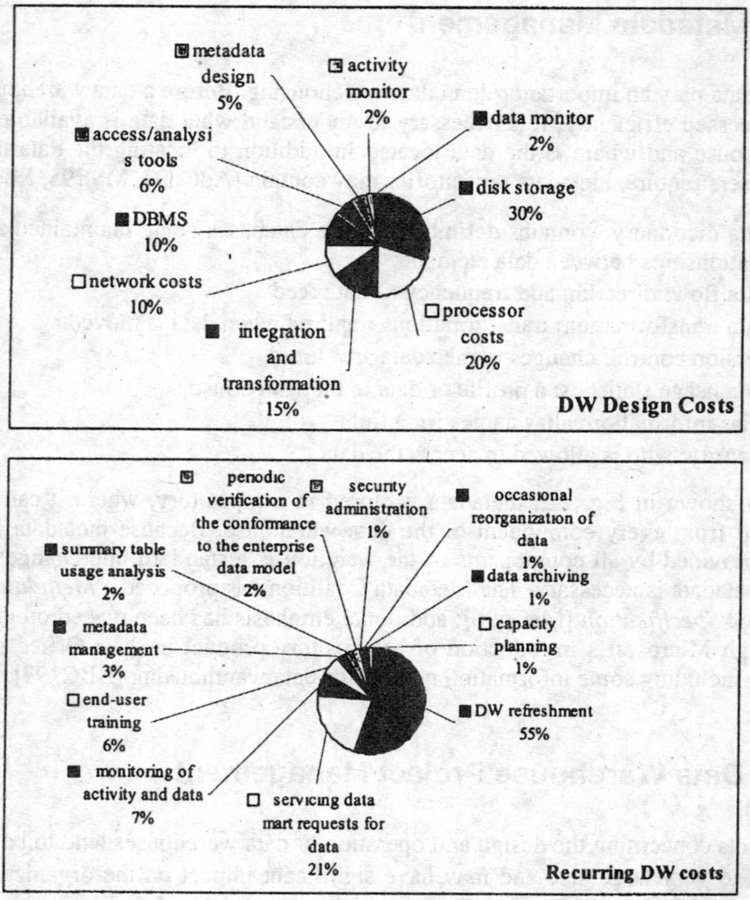

Fig. 1.9. Distribution of data warehouse project costs [Inmo97]

Procedural factors involve deficiencies concerning the deployment of the data warehouse. Apart from classical problems in information systems (IS) management, it is important to notice that end-users must be trained on the new technologies and included in the design of the warehouse. As for the *sociotechnical issues*, the data warehouse may reorganize the way the organization works and intrude the functional or subjective domain of the stakeholders. First, imposing a particular client tool invades the users' desktop, which is considered to be their personal "territory." Second, data ownership is power within an organization. Any attempt to share or take control over somebody else's data is equal to loss of power. Further, no division or department can claim to possess 100% clean, error-free data; revealing the data quality problems within the information system of the department can be frustrating for the affected stakeholders. Fourth, no user community seems to be really willing to shift from gut feeling or experience to objective, data driven management. Last but not least, *ethical considerations* such as privacy concerns need to be taken seriously.

2 Data Warehouse Research: Issues and Projects

In the previous chapter, we have given a broad-brush state of the practice in data warehousing. In this chapter, we look again at the issues of data integration, multidimensional aggregation, query optimization, and data warehouse refreshment, focusing, however, on problems rather than solutions. Each of the topics we address is covered in the following chapters. We give special attention to data warehouse quality management, an issue that is the backbone of all the chapters of the book. Moreover, we briefly review some larger research projects that have addressed more than one of the issues and will therefore be cited in several places throughout the book. Finally, we take a critical overall look at this work and introduce the DWQ conceptual framework which takes the business perspective of data warehousing into account as well as the so far dominant technical aspects.

2.1 Data Extraction and Reconciliation

Data extraction and reconciliation are still carried out on a largely intuitive basis in real applications. Existing automated tools do not offer choices in the quality of service. It is a common phenomenon for the integration process not to be sufficiently detailed and documented; thus making the decisions taken for the integration process difficult to understand and evaluate. We need coherent methodological and tool-based support for the integration activity. The idea of declaratively specifying and storing integration knowledge can be of special importance for supporting high-quality incremental integration and for making all relevant metadata available.

Data reconciliation is first a *source integration* task at the schema level, similar to the traditional task of view integration, but with a richer integration language and therefore with more opportunities for checking the consistency and completeness of data. Wrappers, loaders, and mediators based on such enriched source integration facilities facilitate the arduous task of instance-level *data integration* from the sources to the warehouse, such that a larger portion of inconsistencies, incompatibilities, and missing information can be detected automatically.

2.2 Data Aggregation and Customization

The key purpose of a data warehouse is to support online analytical processing (OLAP), the functional and performance requirements of which are quite different from those of the online transaction processing (OLTP) applications traditionally supported by the operational databases. To facilitate complex analyzes and visualization, the data in a warehouse are organized according to the multidimensional data model. Multidimensional data modeling means the partial aggregation of warehouse data under many different criteria. Usually, the aggregation is performed with respect to predefined hierarchies of aggregation levels.

We need rich schema languages that allow the hierarchical representation of time, space, and numerical/financial domains as well as aggregates over these domains. "Aggregation" means here a grouping of data by some criteria, followed by application of a computational function (sum, average, spline, trend, etc.) for each group. Results on the computational complexity of these language extensions need to be obtained, and practical algorithms for reasoning about metadata expressed in these languages need to be developed and demonstrated. The gain from such research results is better design-time analysis and rapid adaptability of data warehouses, thus promoting the quality goals of relevance, access to nonvolatile historical data, and improved consistency and completeness.

The semantic enrichment will not only enhance data warehouse design but can also be used to optimize data warehouse operation by providing reasoning facilities for semantic query optimization (improving accessibility) and for more precise and better controlled incremental change propagation (improving timeliness). While basic mechanisms stem mostly from active databases, reasoning and optimization techniques on top of these basic services can also use reasoning techniques from Artificial Intelligence (AI) together with quantitative database design knowledge.

2.3 Query Optimization

Data warehouses provide challenges to existing query processing technology for a number of reasons. Typical queries require costly aggregation over huge sets of data, while, at the same time, OLAP users pose many queries and expect short response times; users who explore the information content of a data warehouse apply sophisticated strategies ("drill down," "roll up") and demand query modes like hypothetical ("what if") and imprecise ("fuzzy") querying that are beyond the capabilities of SQL-based systems.

Commercial approaches fail to make use of the semantic structure of the data in a warehouse but concentrate on parallelism or make heavy use of traditional optimization techniques such as indexes or choosing low cost access paths. As a support for OLAP, intermediate aggregate results are precomputed and stored as views.

There are two kinds of metaknowledge in the data warehouse that are relevant integrity constraints expressed in rich schema languages and knowledge about redundancies in the way information is stored. Optimization for nested queries with aggregates can be achieved through the transformation of a query in an equivalent one that is cheaper to compute. Techniques for constraint pushing prune the set of data to be considered for aggregation. Integrity constraints can be used to establish the equivalence of queries that are not syntactically similar. Rewriting techniques reformulate queries in such a way that materialized views are used instead of recomputing previous results. To accomplish its task for queries with aggregation, the query optimizer must be capable to reason about complex relationships between the groupings over which the aggregation takes place. Finally, these basic techniques must be embedded into complex strategies to support OLAP users that formulate many related queries and apply advanced query modes.

2.4 Update Propagation

Updates to information sources need to be controlled with respect to the integrity constraints specified during the design of the data warehouse and derived views. A constraint may state conditions that the data in an information source must satisfy to be of a quality relevant to the data warehouse. A constraint may also express conditions over several information sources that help to resolve conflicts during the extraction and integration process. Thus, an update in a source database may degrade the quality of information, thus resulting in the evolution of the view the data warehouse has over this specific source. It is very important that violations of constraints are handled appropriately, e.g., by sending messages or creating alternative time-stamped versions of the updated data.

Updates that meet the quality requirements defined by integrity constraints must then be propagated towards the views defined at the data warehouse and user level. This propagation must be done efficiently in an incremental fashion. Recomputation can take advantage of useful cached views that record intermediate results. The decision to create such views depends on an analysis of both the data warehouse query workload and the update activity at the information sources. This requires the definition of new design optimization algorithms that take into account these activities and the characteristics of the constraints that are considered for optimization.

Integrity constraints across heterogeneous information sources are currently managed manually, resulting in unreliable data. At best, a common reference data model enforces the satisfaction of some structural and quality constraints. However, this approach is quite rigid and does not enable an easy integration of new information sources. Another approach suggests a flexible and declarative definition of constraints during the design phase. Constraints can then be mapped to active rules, which are increasingly accepted as a suitable base mechanism for enforcing constraints.

2.5 Modeling and Measuring Data Warehouse Quality

Although many data warehouses have already been built, there is no common methodology that supports database system administrators in designing and evolving a data warehouse. The problem with architecture models for data warehouses is that practice has preceded research in this area and continues to do so. Consequently, the task of providing an abstract model of the architecture becomes more difficult.

The architecture of a data warehouse addresses questions such as:

- What kinds of components are used in a data warehouse environment?
- How do these components exchange (meta) data?
- How can the quality of a data warehouse be evaluated and designed?

Formally, an architecture model corresponds to the schema structure of the metadatabase that controls the usually distributed and heterogeneous set of data warehouse components and therefore is the essential starting point for design and operational optimization. Expressiveness and services of the metadata schema are crucial for data warehouse quality. The purpose of architecture models is to provide an expressive, semantically defined, and computationally understood meta-modeling language based on existing approaches in practice and research.

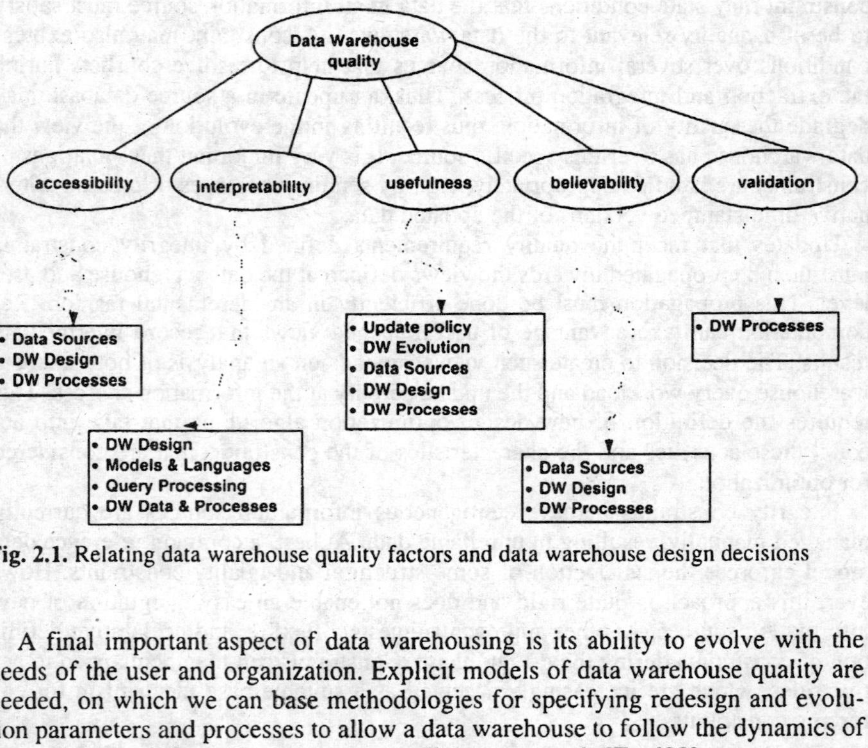

Fig. 2.1. Relating data warehouse quality factors and data warehouse design decisions

A final important aspect of data warehousing is its ability to evolve with the needs of the user and organization. Explicit models of data warehouse quality are needed, on which we can base methodologies for specifying redesign and evolution parameters and processes to allow a data warehouse to follow the dynamics of the underlying information sources as well the user needs [Engl99].

Figure 2.1 gives a rough overview of how specific tasks in data warehouse development and operation can be linked to typical data quality goals such as

- Accessibility: how to make data more accessible to users.
- Interpretability: how to help users understand the data they get.
- Usefulness: how to fit access to the data warehouse into the users' work processes and give them the data they need.
- Believability: how to make the user trust the data, and how to make data from possibly unreliable sources more trustworthy.
- Validation: how to ensure that all the above quality issues have actually been adequately addressed.

The static architecture, integration, and aggregation modeling languages should be augmented with evolution operators which support controlled change in the data warehouse structure. This includes the addition, deletion, or reevaluation of information sources, the efficient migration between different structures of derived aggregated data, and the tracing of such changes across the data warehouse architecture. Taken together, these evolution tools evolve the vision of a data warehouse as a passive buffer between operational and analytic processing towards that of a competitive information supermarket that carefully selects and prepares its products, packages and displays them nicely, and continuously adapts to customer demands in a cost-efficient manner.

The above operators provide the means to control and trace the process of evolving a data warehouse application. However, many steps involve options such as what information sources to use, what update policies to select, and what views to materialize. A control center for data warehouse planning may help the data warehouse manager to quantitatively evaluate these options and find an optimal balance among the different policies involved, prior to executing evolution steps. In particular, it may enable the identification of policies for optimized view materialization based on workload parameters (sets of expected queries and updates). Moreover, enriched metadata representation and reasoning facilities can be used in data warehouse design to provide rigorous means of controlling and achieving evolution. Such reasoning mechanisms can help determine whether certain information sources can substitute each other and whether certain materialized views can be reused to materialize others cheaply. They will significantly change the cost trade-offs, making evolution easier than currently feasible.

2.6 Some Major Research Projects in Data Warehousing

To summarize, we need techniques and tools to support the rigorous design and operation of data warehouses

- Based on well-defined data quality factors
- Addressed by a rich semantic approach
- Realized by bringing together enabling technologies

In the presence of such facilities, the data warehouse development time for a given quality level can be significantly reduced and adaptation to changing user demands can be further facilitated. There is already a high demand for design tools for distributed databases which are not satisfied by current products.

Regrettably, the coverage of existing research projects does not address all the questions either. Most research in data warehousing focuses on source integration and update propagation. We sketch the approaches of several well-known recent projects: the Information Manifold (IM) developed at AT&T, the TSIMMIS) project at Stanford University, the Squirrel project at the University of Colorado, and the WHIPS project at Stanford University.

The **Information Manifold** (IM) system was developed at AT&T for information gathering from disparate sources such as databases, SGML documents, and unstructured files [LeSK95, KLSS95, LeRO96]. It is based on a rich domain model, expressed in a knowledge base, which allows for describing various properties of the information sources from the topics they are about to the physical characteristics they have. This enables users to pose high-level queries to extract the information from different sources in a unified way. The architecture of IM suits the dynamic nature of the information sources. In particular, to add a new source, only its description (i.e., its view as a relational schema and the related types and constraints) needs to be added, without changing the existing integration mechanisms of the system. IM does not allow any form of update in either the information sources or the world view. Updates on the information sources are external to IM, while propagation of updates from the world view to the single sources is not supported.

The Stanford-IBM Manager of Multiple Information Sources (TSIMMIS) was a project that shares with IM the goal of providing tools for the integrated access to multiple and diverse information sources and repositories [CGH*94, Ullm97]. Each information source is equipped with a *wrapper* that encapsulates the source, converting the underlying data objects to a common simple object-oriented data model called *Object Exchange Model* (OEM). On top of wrappers TSIMMIS has another kind of system components: the *mediators* [Wied92]. Each mediator obtains information from one or more wrappers or other mediators, refines this information by integrating and resolving conflicts among the pieces of information from the different sources, and provides the resulting information to the user or other mediators.

The TSIMMIS query language is a SQL-like language adapted to treat OEM objects. When a user poses a query to the system, a specific mediator is selected. Such a mediator decomposes the query and propagates the resulting sub-queries to the levels below it (either wrappers or mediators). The answer provided by such levels is then reconstructed performing integration steps and processing, using off-the-shelf techniques.

Squirrel was a prototype system at the University of Colorado [ZHKF95, ZHKF95a, HuZh96, ZhHK96] which provides a framework for data integration based on the notion of an *integration mediator*. Integration mediators are active modules that support integrated views over multiple databases. A Squirrel mediator consists of a query processor, an incremental update processor, a virtual attribute processor, and a storage system to store the materialized views. These mediators are generated from high-level specifications. In a mediator, a view can be fully materialized, partially materialized, or fully virtual.

The queries which are sent to the mediator are processed by the query processor using the materialized view or by accessing the source databases, if the necessary information is not stored in the mediator. The update processor maintains the materialized views incrementally using the incremental updates of the sources. The virtual update processor accesses the sources if the information to answer a query is not available in the mediator.

The architecture of a Squirrel mediator consists of three components: a set of active rules, an execution model for such rules, and a View Decomposition Plan (VDP). The notion of VDP is analogous to the query decomposition plan in query optimization. More specifically, the VDP specifies the classes that the mediator maintains, and provides the basic structure for supporting incremental maintenance.

The **WareHouse Information Prototype at Stanford (WHIPS)** project [HGW*95, WGL*96] developed a data warehouse prototype test bed to study algorithms for the collection, integration, and maintenance of information from heterogeneous and autonomous sources. The WHIPS architecture [WGL*96] consists of a set of independent modules implemented as CORBA objects to ensure modularity and scalability. The central component of the system is the *integrator* to which all other modules report. Different data models may be used both for each source and the data warehouse data. The relational model is used as a unifying model, and for each source and the data warehouse the underlying data are converted to the relational model by a specific *wrapper*.

2.7 Three Perspectives of Data Warehouse Metadata

Almost all current research and practice understands data warehouse architecture as a stepwise information flow from information sources through materialized views towards analyst clients, as shown in Fig. 2.2. For example, projects such as TSIMMIS [CGH*94], Squirrel [HuZh96], and WHIPS [HGW*95] all focus on the integration of heterogeneous data via wrappers and mediators, using different logical formalisms and technical implementation techniques. The IM project [LeSK95] is the only one providing a conceptual domain model as a basis for integration.

In the following, we show that this architecture neglects the business role of data warehouses and describe an extended architecture and data warehouse metamodel realized in the DWQ project and intended to address this problem.

DWQ [JaVa97, DWQ99] was a European Esprit project involving three universities (RWTH Aachen/Germany, NTUA Athens/Greece, and Rome La Sapienza/Italy) and three research centers (DFKI/Germany, INRIA/France, and IRST/Italy) to investigate the Foundations of Data Warehouse Quality. The basic goal was to enrich the semantics of data warehouse modeling formalisms to improve several aspects of data warehouse design and operation. Major topics include the definition of an extended metamodel for data warehouse architecture and quality [JJQV99] inference techniques for improving source integration [CDL*98] as well as working with multidimensional data models [BaSa98, GeJJ97, Vass98], systematic design of refreshment policies for data warehousing [BFL*97], and optimizing the choice of materialized views [ThSe97].

The traditional data warehouse architecture, advocated both in research and in the commercial trade press, is recalled in Fig. 2.2. Physically, a DW system consists of databases (source databases, materialized views in the distributed data warehouse), data transport agents that ship data from one database to another, and a data warehouse repository which stores metadata about the system and its evolution. In this architecture, heterogeneous information sources are first made accessible in a uniform way through extraction mechanisms called *wrappers*, then *mediators* [Wied92] take on the task of information integration and conflict resolution. The resulting standardized and integrated data are stored as materialized views in the data warehouse. The DW base views are usually just slightly aggregated; to customize them better for different groups of analyst users, *data marts* with more aggregated data about specific domains of interest are frequently constructed as second-level caches which are then accessed by data analysis tools ranging from query facilities through spreadsheet tools to full-fledged data mining systems that employ knowledge-based or neural network techniques.

The content of the repository (metadatabase) determines to a large extent the usage and evolution of the data warehouse. The main goal of the DWQ approach is therefore to define a metadatabase structure which can capture and link all relevant aspects of DW architecture, process, and quality.

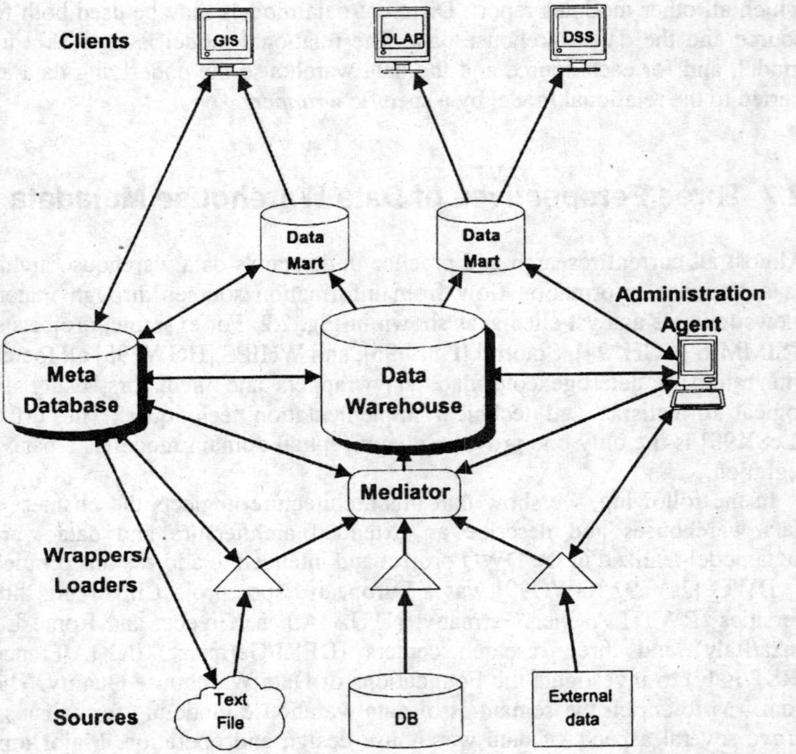

Fig. 2.2. Traditional data warehouse architecture

Our key observation is that the architecture in Fig. 2.2 covers only partially the tasks faced in data warehousing and is therefore unable to even express, let alone support, a large number of important quality problems and management strategies.

The main argument we wish to make is the need for a *conceptual enterprise perspective*. To explain, consider Fig. 2.3 where the traditional flow of information is stylized on the right-hand side, whereas the process of creating and using the information is shown on the left. Suppose an analyst wants to know something about the business – the question mark in Fig. 2.3. The analyst does not have the time to observe the business directly but must rely on existing information gained by operational departments, and documented as a side effect of OLTP systems. This way of gathering information implies already a bias which needs to be compensated when selecting OLTP data for uploading and cleaning into a DW where it is then further preprocessed and aggregated in data marts for certain analysis tasks. Considering the long path the data has taken, it is obvious that also the last step, the formulation of conceptually adequate queries and the conceptually adequate interpretation of the answers presents a major problem to the analyst.

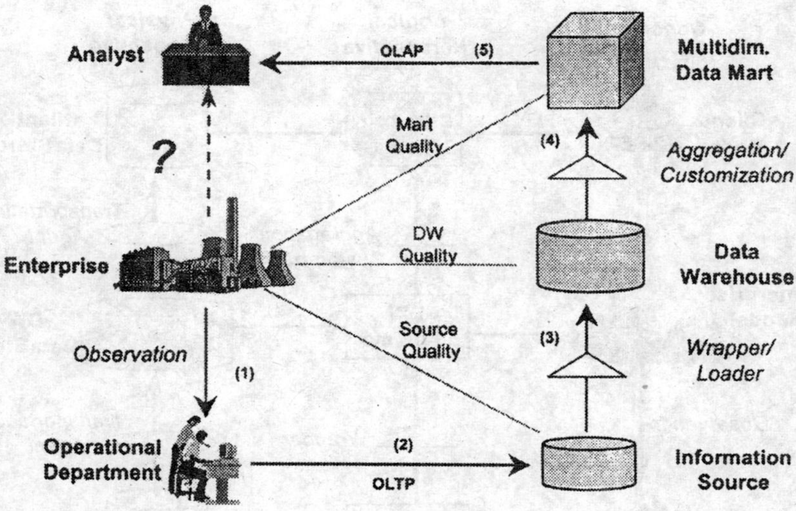

Fig. 2.3. Data warehousing in the context of an enterprise

Indeed, Fig. 2.2 only covers two of the five steps shown in Fig. 2.3. Thus, it has no answers to typical practitioner questions such as "*How come my operational departments put so much money in their data quality, and still the quality of my DW is terrible?*" (answer: the enterprise views of the operational departments are not easily compatible with each other or with the analysts view), or "*What is the effort required to analyze problem X for which the DW currently offers no information?*" (could simply be a problem of wrong aggregation in the materialized views, could require access to not-yet-integrated OLTP sources, or could even involve setting up new OLTP sensors in the organization).

An adequate answer to such questions requires an explicit model of the conceptual relationships among an enterprise model, the information captured by OLTP departments, and the OLAP clients whose task is the decision analysis. We have argued that a DW is a major investment undertaken for a particular business purpose. We therefore do not just introduce the enterprise model as a minor part of the environment, but demand that *all other models are defined as views on this enterprise model*. Perhaps surprisingly, even information source schemas define views on the enterprise model – not vice versa as suggested by Fig. 2.2 !

By introducing an explicit business perspective as in Fig. 2.3, the wrapping and aggregation transformations performed in the traditional data warehouse literature can thus all be checked for interpretability, consistency or completeness with respect to the enterprise model – provided an adequately powerful representation and reasoning mechanism is available.

At the same time, the logical transformations need to be implemented safely and efficiently by physical data storage and transportation – the third perspective in our approach. It is clear that these physical aspects require completely different modeling formalisms from the conceptual ones. Typical techniques stem from queuing theory and combinatorial optimization.

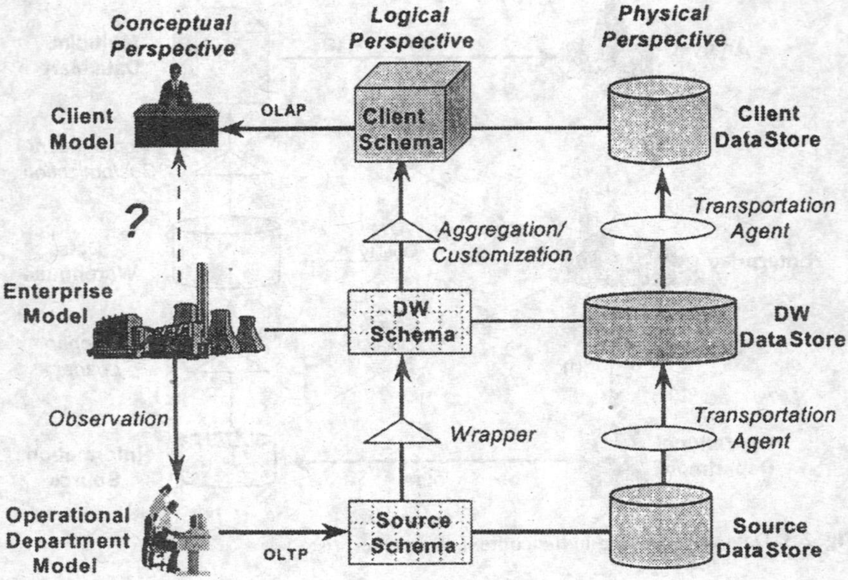

Fig. 2.4. The proposed data warehouse metadata framework

As a consequence, the data warehouse metadata framework we propose in Fig. 2.4 clearly separates three perspectives: a conceptual enterprise perspective, a logical data modeling perspective, and a physical data flow perspective.

As shown in Fig. 2.5, this framework can be instantiated by information models (conceptual, logical, and physical schemas) of particular data warehousing strategies which can then be used to design and administer the instances of these data warehouses – the main role of administration and metadatabase in Fig. 2.2.

However, quality cannot just be assessed on the network of nine combinations of perspectives and levels but is largely determined by the processes how these are constructed. The process meta model defines a way how such processes can be defined, the process models define plans how data warehouse construction and administration are to be done, and the traces of executing such plans are captured at the lowest level. This process hierarchy accompanying the DW product model is shown on the right of Fig. 2.5.

In the following chapters of this book, we keep this basic framework in mind when we present, and sometimes criticize, the state of the practice and state of the art in the four key subprocesses of data warehousing: integration of new data sources, operational data refreshment, mapping from traditional to multidimensional data, and query processing. Finally, in Chaps. 7 and 8, we return to the metadata framework and show how its implementation in an extended repository can be used for data warehouse quality analysis and quality-driven data warehouse design.

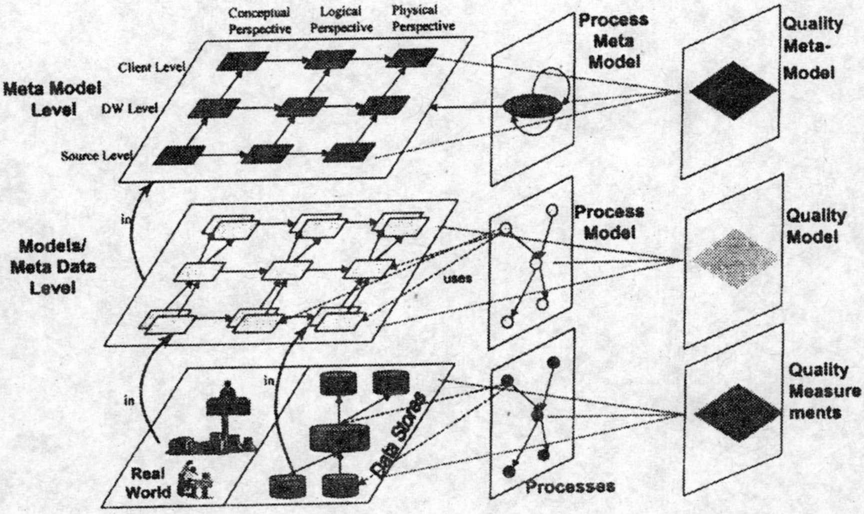

Fig. 2.5. Repository structure capturing product, process, and quality of data warehousing

3 Source Integration

According to [Inmo96], integration is the most important aspect of a data warehouse. When data passes from the application-oriented operational environment to the data warehouse, possible inconsistencies and redundancies should be resolved, so that the warehouse is able to provide an integrated and reconciled view of data of the organization.

The type of integration to be carried out in the design of a data warehouse can be seen within the field of information integration, which has been investigated in different areas such as:

- Databases: where both schema and data integration are central issues;
- Cooperative information systems: where integration is one of the problems arising when component information systems cooperate in order to perform their tasks, for example by sharing data;
- Global information systems: constituted by heterogeneous information sources that are put online, e.g., on the Web, and accessed in an integrated way;
- Knowledge representation: where typically integration is considered in the more general context of conforming and merging logical theories or other kinds of knowledge expressed in a more general form than just data.

Here, we deal with the first area, which is the appropriate one to look at, in the context of data warehouses. The chapter is organized as follows. In Sect. 3.1 we take a look at the state of practice in source integration, by briefly describing tools and products that have been designed for supporting such a task. In Sect. 3.2 we review relevant research work in the context of schema and data integration. Finally, in Sect. 3.3 we discuss the principles for devising a general systematic methodology for source integration in data warehousing.

3.1 The Practice of Source Integration

The market offers a large variety of methods, tools, and products under the label "Data Warehouse." To address the portion of the market that is concerned with source integration some remarks are in order, concerning the typology of the products being sold and the methodologies that are proposed in connection with them.

The problem of data warehouse construction is being specifically addressed by monolithic tools centered on the service of data warehouse management, as well as by tool suites that collect products which are marketed as data warehouse *solutions*. The latter can be provided by the same vendor but can also be put together to include components from third parties, which are available as *separate* products. The way the various components of such solutions are organized varies, and it is very difficult to isolate tools that are specific to source integration.

Since the problem of the data warehouse construction is closely related to the problem of database construction, the methodologies that are proposed in this context have their roots in the tradition of database design. With regards to source integration, the most significant feature required for data warehouse construction is the ability to deal with several information sources. Basically, two main approaches have been proposed:

1. Construction of an integrated enterprise schema (often called enterprise model)
2. Focus on explicit mappings between the data warehouse and the sources

The first approach is proposed by those vendors who have built tools and associated methodologies for enterprise modeling. Without entering in a detailed analysis of the methodologies associated with specific tools, we remark that the most extensively supported model is the Entity-Relationship model, which can sometimes be derived through reverse engineering tools. In addition, some of the products can deal with the star and snowflake models, but there is little support to model transformations from standard models to data warehouse specific ones. For a more detailed description of these aspects, we refer the reader to Chap. 1.

With respect to the second approach, typically the mapping between the data warehouse and the sources is done at the logical level by relating the relations of the data warehouse to those of the source, sometimes taking into consideration the actual code for extracting the data from the source. To this purpose, graphical and structure directed editors guide the user in the specification of the relationship between the data warehouse and the information sources. However, there is no support for the verification of properties of the specified relationships or properties that can be derived from them.

To identify the market products that are specific to the task of source integration, we distinguish between the tasks of schema and data integration. For the former, we look both at products marketed as data warehouse management systems and at multicomponent data warehouse solutions, trying to highlight the aspects that are specific to schema integration. For the latter, we address the products for data integration, including data extraction, data cleaning, and data reconciliation, which are more easily identifiable as separate tools or separate components. For a more extensive description of such products, the interested reader is referred to Chap. 4.

3.1.1 Tools for Data Warehouse Management

The warehouse manager is either the central component of a data warehouse solution or a complex product including several functionalities needed for the construction of a data warehouse. The market offers a great variety of proposals that differ with respect to the role played by the manager. Clearly, there is the need to provide a common basis for both the data warehouse construction process and the access and analysis of data. But stronger attention is paid to the first aspect, since most of the tools for data access and analysis rely on a standard database system as the enterprise data warehouse (as long as it provides enough capacity of data storage and handling). The main function of the warehouse manager is to handle the metadata (i.e., data schemata). Consequently, a large portion of the proposals for data warehouse management arise from the companies that have been supporting tools and methodologies for data modeling. Among the metadata we find a description of the information sources, of the data warehouse itself and of the mapping between the two. As previously remarked, one possibility is to build the data warehouse by constructing an integrated enterprise schema, in which case the system supports both the conceptual modeling of the data and the schema integration process. The possibility of allowing integration through an explicit mapping of the relationships between the logical schema of the data warehouse and the sources is described in the next subsection.

In addition, the warehouse manager is responsible for handling the extraction, update, and maintenance processes. So, apart from the previously mentioned logical specification of the data extraction process, a warehouse manager allows the user to specify a scheduling for the extraction of data and update policies. This functionality can either be automatically enforced from a high-level user specification, typically based on a set of predefined choices, or must be explicitly coded by the user. This aspect is further addressed in Chap. 4.

Another functionality that is sometimes supported by the warehouse manager is the connection to the tools used for data analysis. In this respect, the warehouse manager can handle metadata about the data organization used by these tools. Clearly, this is important to keep a connection between the models used for data analysis and the models of the data in the warehouse and, consequently, in the information sources. In particular, this functionality makes it possible to provide a unified framework for handling the design and transformation of the various data models needed in data warehousing.

Among the tools that support data warehouse management we can identify a first group which is rooted on conventional data modeling and supports reverse engineering and schema integration. Among them we find Groundworks of Cayenne (Bachman), ERWIN ERX of LogicWorks, and Kismetic Analyst of Kismeta. A second group of solutions is more focused on the construction and maintenance of the data warehouse, among them the Warehouse Manager of Prism Solutions, SAS Warehouse Manager of SAS Institute, the Metacube Warehouse Manager of Informix, and Sourcepoint of Software AG.

3.1.2 Tools for Data Integration

The tools that are used in the process of data integration specifically support the extraction of data from the sources and, consequently, provide both the interface with the operational databases and the mechanisms for the transformation of data that make them suitable for the data warehouse.

With respect to the extraction of data, the products may be divided in software development environments, whose capabilities range from automatic software generation to more traditional programming environments to query handlers which can deal with a variety of query languages, typically SQL dialects.

In the first case, the user is required to provide, typically through an interactive session as hinted above, the mapping between the data to be extracted from the information sources and the form they take in the data warehouse. Otherwise, the procedures for extracting the data from the sources need to be hand coded. It is worth noting that if the mapping from the sources to the target is not a one-to-one mapping, the possibility of adding hand-coded extraction procedures is left as the only possible way out.

Several forms of data transformation and data cleaning can be associated with data extraction:

- *Data reconciliation*: integration of different data formats, codes, and values, based on predefined mappings
- *Data validation*: identification of potentially inconsistent data that can be either removed or fixed
- *Data filtering*: filtering of the data according to the requirements of the data warehouse

In addition, data mining is sometimes suggested at this stage, thus avoiding the transfer of large quantities of data to the tools for data analysis.

The products for data transformation can either filter the data while they are loaded in the data warehouse or perform a database auditing function after the data have been loaded. Often, several kinds of data transformation can be supported by a single tool. In general, the tools for data transformation and cleaning are quite successful, although they are rather limited in their scope, since they are restricted to specific kinds of data, e.g., postal codes, addresses, etc.

Here is a list of tools that fall into this category: Carleton PassPort of Carleton Corporation, EXTRACT of Evolutionary Tech., Metacube Agents of Informix, Sybase Mpp, Datajoiner of IBM. In addition, The Warehouse Manager of Prism Solutions and SAS Warehouse Manager of SAS Institute include internal facilities for the tasks of data integration. Finally, among the tools for data mining we have Integrity of Vality Tech. and Wizrule of Wizsoft.

3.2 Research in Source Integration

A large body of work has been carried out in recent years in database integration. We classify the various approaches with regards to the context of integration, the kind of inputs and outputs of the integration process, and the goal of the process itself:

1. *Schema integration.* In this case, the input of the integration process is a set of (source) schemata, and the output is a single (target) schema representing the reconciled intentional representation of all input schemata. The output also includes the specification of how to map source data schemata into portions of the target schema.
2. *Virtual data integration.* The input is a set of source data sets, and the output is a specification of how to provide a global and unified access to the sources to satisfy certain information needs, without interfering with the autonomy of the sources.
3. *Materialized data integration.* As in the previous case, the input is a set of source data sets, but here the output is a data set representing a reconciled view of the input sources, both at the intentional and the extensional level.

Schema integration is meaningful, for example, in database design and, in general, in all the situations where descriptions of the intentional level of information sources are to be compared and unified. Virtual data integration has been adopted in several multidatabase and distributed database projects. Materialized data integration is most directly meaningful in information system reengineering and data warehousing, but the other approaches are required to support it as well.

Each of the three classes of techniques is associated with specific questions that need to be answered to propose and/or evaluate a solution.

Schema integration must be done, either implicitly or explicitly, to integrate data. This makes work on schema integration relevant to any data integration approach and in particular to the context of data warehousing. In reviewing relevant work in the context of schema integration, we make use of several key aspects, which can be summarized as follows:

1. Is a global schema produced or not?
2. Which is the methodological step in the source integration process the work refers to (preintegration, schema comparison, schema conforming, schema merging and restructuring)?
3. Which is the formalism for representing schemata?
4. To what extent is the notion of schema quality taken into account (correctness, completeness, minimality, understandability [BaLN86])?

As previously mentioned, data integration, besides comparing and integrating source schemata, also deals with the additional problem of merging actual data stored in the sources. Therefore, one of the aspects that distinguishes data integration from schema integration is *object matching*, i.e., establishing when different objects in different source databases represent in fact the same real-world element and should therefore be mapped to the same data warehouse object. Such an aspect has a strong impact on data quality dimensions like redundancy and accuracy. The simplest object matching criterion, called *key-based criterion*, is to identify objects having the same key. Approaches based on lookup-tables and identity functions use more complex criteria to match objects.

Albert [Albe96] argues that object identity should be an equivalence relation and therefore proposes to represent objects in the integrated schema as equivalence classes. If one wants to respect this requirement, however, the evaluation of queries against the integrated schema may become more complex. In fact, in the presence of subtype relationships, even simple nonrecursive queries against the integrated schema may have to be translated to recursive queries against the underlying data, reflecting the transitive property of the equivalence relation.

While precise methodologies have been developed for schema integration, there is no consensus on the methodological steps for data integration, with either virtual or materialized views. In reviewing the state of the art in this research area, we use as guiding criteria the various approaches and implementation efforts in data integration. Many of the systems that we will survey have already been discussed in Chap. 1 with reference to the architectural aspects. Here, we concentrate on the aspects that are related to source integration.

Data integration with virtual views and with materialized views is sometimes considered as complementary approaches.

In a *virtual views approach*, data are kept only in the sources and are queried using the views. Hence much emphasis is put on query decomposition, shipping, and reconstruction: a query needs to be decomposed in parts, each of which is shipped to the corresponding source. The data returned are then integrated to reconstruct the answer to the query.

When discussing papers dealing with virtual data integration, we consider the following additional points:

1. Is a global view assumed or not?
2. Which is the methodological step in the virtual integration process the work refers to (query decomposition, shipping, reconstruction)?
3. Which is the formalism for representing data (file, legacy, relational DB, OODB, unstructured)?
4. Which query language is used for posing global queries?
5. Which data matching criteria are adopted (key-based, lookup table-based, comparison-based, historical-based)?
6. To what extent is the notion of data quality taken into account (interpretability, usefulness, accessibility, credibility, etc.)?

In a *materialized views approach*, views are populated with data extracted from the sources. Query decomposition, shipping, and reconstruction are used in the phase of populating views rather than in the process of query answering. Additionally, issues such as maintenance strategies and scheduling that are absent in the context of virtual views become crucial for materialized views. We remark that most of the issues that arise in the approaches based on virtual views are of importance for materialized views approaches as well. So, despite differences in focus, we take the position that materialized views are conceptually a "specialization" of virtual views.

When reviewing papers dealing with materialized data integration, the following aspects are further taken into account (partly borrowed from [HuZh96]):

1. Which data are materialized?
2. What is the level of activeness of the sources (sufficient, restricted, nonactive)?
3. What is the maintenance strategy (local incremental update, polling-based, complete refresh) and timing?

3.2.1 Schema Integration

Schema integration is the activity of integrating the schemata of the various sources to produce a homogeneous description of the data of interest. This activity is traditionally performed in a *one-shot* fashion, resulting in a global schema in which all data are represented uniformly [BaLN86]. More recently, to deal with autonomous and dynamic information sources, an incremental approach is arising [CaLe93]. Such an approach consists of building a collection of independent partial schemata, formalizing the relationships among entities in the partial schemata by means of so-called *interschema assertions*. In principle, under the assumption that the schemata of the various information sources remain unchanged, the incremental approach would eventually result in a global schema, similar to those obtained through a traditional one-shot approach, although in practice, due to the dynamics of the sources, such a result is never achieved. Additionally, the integration may be partial, taking into account only certain aspects or components of the sources [CaLe93].

Independently of the integration strategy adopted, schema integration is divided into several methodological steps, with the aim of relating the different components of schemata, finding and resolving conflicts in the representation of the same data among the different schemata, and eventually merging the conformed schemata into a global one. In [BaLN86], the following methodological steps are singled out:

- Preintegration
- Schema comparison
- Schema conforming
- Schema merging and restructuring.

We review recent studies on schema integration, according to the steps they address. We refer the interested reader to [BaLN86] for a comprehensive survey on previous work in this area.

3.2.1.1 Preintegration

Preintegration consists of an analysis of the schemata to decide the general integration policy: choosing the schemata to be integrated, deciding the order of integration, and possibly assigning preferences to entire schemata or portions thereof. The choices made in this phase influence the usefulness and relevance of the data corresponding to the global schema. During this phase additional information relevant to integration is also collected, such as assertions or constraints among views in a schema. Such a process is sometimes referred to as *semantic enrichment* [GaSC95, GaSC95a, BlIG94, RPRG94]. It is usually performed by translating the source schemata into a richer data model, which allows for representing information about dependencies, null values, and other semantic properties, thus increasing interpretability and believability of the source data.

For example, Blanco et al. [BlIG94] enrich relational schemata using a class-based logical formalism, *description logic* (DL), available in the terminological system BACK [Pelt91]. In a different approach, Garcia-Solaco et al. [GaSC95, GaSC95a] use as unifying model a specific object-oriented model with different types of specialization and aggregation constructs.

Sheth et al. [ShGN93] propose the creation of a knowledge base (*terminology*) in the preintegration step. More precisely, a hierarchy of attributes is generated, representing the relationship among attributes in different schemata. Source schemata are classified, the terminology thus obtained corresponds to a partially integrated schema. Such a terminology is then restructured by using typical reasoning services of class-based logical formalisms. The underlying data model is hence the formalism used for expressing the terminology, more precisely, a description logic (called CANDIDE) is used.

Johanneson [Joha94] defines a collection of transformations on schemata represented in a first order language augmented with rules to express constraints. Such transformations are correct with respect to a given notion of information preservation, and constitute the core of a "standardization" step in a new schema integration methodology. This step is performed before schema comparison and logically subsumes the schema conforming phase, which is not necessary in the new methodology.

3.2.1.2 Schema Comparison

Schema comparison consists of an analysis to determine the correlations among concepts of different schemata and to detect possible conflicts. Interschema properties are typically discovered during this phase.

In [BoCo90], schema comparison in an extended entity-relationship model is performed by analyzing structural analogies between subschemata through the use of similarity vectors. Subsequent conforming is achieved by transforming the structures into a canonical form.

The types of conflicts that arise when comparing source schema components have been studied extensively in the literature (see [BaLN86, KrLK91, SpPD92, OuNa94, RPRG94]) and consensus has arisen on their classification, which can be summarized as follows:

- *Heterogeneity conflicts* arise when different data models are used for the source schemata.
- *Naming conflicts* arise because different schemata may refer to the same data using different terminologies. Typically one distinguishes between *homonyms*, where the same name is used to denote two different concepts, and *synonyms*, where the same concept is denoted by different names.
- *Semantic conflicts* arise due to different choices in the level of abstraction when modeling similar real-world entities.
- *Structural conflicts* arise due to different choices of constructs for representing the same concepts.

In general, this phase requires a strong knowledge of the semantics underlying the concepts represented by the schemata. The more semantics is represented formally in the schema, the easier similar concepts in different schemata can be automatically detected, possibly with the help of specific CASE tools that support the designer. Traditionally, schema comparison was performed manually [BaLN86]. However recent methodologies and techniques emphasize automatic support to this phase.

For example, Gotthard et al. [GoLN92] propose an architecture where schema comparison and the subsequent phase of schema conforming are iterated. At each cycle the system proposes correspondences between concepts that can be confirmed or rejected by the designer. Newly established correspondences are used by the system to conform the schemata and to guide its proposals in the following cycle. Both the component schemata and the resulting global schema are expressed in a data model that essentially corresponds to an entity-relationship model extended with complex objects.

Blanco et al. [BlIG94] exploit the reasoning capabilities of a terminological system to classify relational schema components and derive candidate correspondences between them expressed in the description logic BACK.

Miller et al. [MiYR94] study the problem of deciding equivalence and dominance between schemata, based on a formal notion of information capacity given in [Hull86]. The schemata are expressed in a graph-based data model which allows the representation of inheritance and simple forms of integrity constraints. First, it is shown that such a problem is undecidable in schemata that occur in practice. Then *sufficient* conditions for schema dominance are defined which are based on a set of schema transformations that preserve schema dominance. A schema S1 is dominated by a schema S2 if there is a sequence of such transformations that converts S1 to S2.

Reconciliation of semantic discrepancies in the relational context due to information represented as data in one database and as metadata in another is discussed in [KrLK91]. The paper proposes a solution based on reifying relations and databases by transforming them into a structured representation.

3.2.1.3 Schema Conforming

The goal of this activity is to conform or align schemata to make them compatible for integration. The most challenging aspect is represented by conflict resolution which in general cannot be fully automated. Typically, semiautomatic solutions to schema conforming are proposed, in which human intervention by the designer is requested by the system when conflicts have to be resolved. Recent methodologies and techniques emphasize also the automatic resolution of specific types of conflicts (e.g., structural conflicts). However, a logical reconstruction of conflict resolution is far from being accomplished and still is an active topic of research.

Qian [Qian96] studies the problem of establishing correctness of schema transformations on a formal basis. More specifically, schemata are modeled as abstract data types, and schema transformations are expressed in terms of signature interpretations. The notion of schema transformation correctness is based on a refinement of Hull's notion of information capacity [Hull86]. In particular, such a refinement allows the formal study of schema transformations between schemata expressed in different data models.

Vidal and Winslett [ViWi94] present a general methodology for schema integration in which the semantics of updates is preserved during the integration process. More precisely, three steps are defined combination, restructuring, and optimization. In the combination phase, a combined schema is generated, which contains all source schemata and assertions (constraints) expressing the relationships among entities in different schemata. The restructuring step is devoted to normalizing (through schema transformations) and merging views, thus obtaining a global schema, which is refined in the optimization phase. Such a methodology is based on a semantic data model which allows the declaration of constraints containing indications on what to do when an update violates that constraint. A set of schema transformations are defined, also being *update semantics preserving*, in the sense that any update specified against the transformed schema has the same effect as if it were specified against the original schema.

3.2.1.4 Schema Merging and Restructuring

During this activity the conformed schemata are superimposed, thus obtaining a (possibly partial) global schema. Such a schema is then tested against quality dimensions such as completeness, correctness, minimality, and understandability. This analysis may give rise to further transformations of the obtained schema.

Buneman et al. [BuDK92] define a technique for schema merging, which consists of a binary merging operator for schemata expressed in a general data model. Such an operator is both commutative and associative; hence, the resulting global schema is independent of the order in which the merges are performed.

Spaccapietra et al. [SpPD92] present a methodology for schema integration which allows the automatic resolution of structural conflicts and the building of the integrated schema without requiring the conformance of the initial schemata. The methodology is applicable to various source data models (relational, entity-relationship, and object-oriented) and is based on an expressive language to state interschema assertions that may involve constructs of schemata expressed in different models. Data model independent integration rules that correspond to the interschema assertions are defined in the general case and are also specialized for the various classical data models. Quality issues are addressed in an informal way. In particular, correctness is achieved by selecting, in case of conflicts, the constructs with weaker constraints. The methodology includes strategies that avoid introducing redundant constructs in the generated schema. Completeness, however, is not guaranteed since the model adopted for the global schema lacks a generalization construct.

Geller et al. [GPNS92, GPC*92] present an integration technique (*structural integration*) which allows for the integration of entities that have structural similarities, even if they differ semantically. An object-oriented model, called the DUAL model, is used, in which structural aspects are represented as object types, and semantic aspects are represented as classes. Two notions of correspondence between classes are defined full structural correspondence and partial structural correspondence. The (partial) integration of two schemata is then obtained through a generalization of the classes representing the original schemata.

3.2.2 Data Integration – Virtual

As previously mentioned, in a *virtual views approach*, data are kept only in the sources and are queried using the views. It follows that the virtual view approach is not fully suited for data warehousing. However, some of the aspects of this approach, such as query decomposition, are relevant in data warehousing and will be analyzed in the following.

3.2.2.1 Carnot

In the *Carnot* system [CoHS91, HJK*93], individual schemata are mapped onto a large ontology which is provided by the CYC knowledge base [LeGu90]. Such an ontology is expressed in an extended first order representation language called *Global Context Language* (GCL). The high expressiveness of GCL allows representation in the global CYC schema of both metamodels for various schema formalisms and all available knowledge about the individual schemata, including integrity constraints, allowed operations, and organizational knowledge. The mapping between each model and the global context involves both syntax and semantics and is stated in terms of so called *articulation axioms* which play the role of interschema assertions. Once the mapping is established, queries are answered and updates are performed by first translating them to GCL and then distributing them to the different resource management systems.

A similar architecture also based on a first order formalism for the unified schema is proposed in [DiWu91]. Source schemata and global views are represented in a knowledge base, and an inference engine based on rules is used to extract integrated information from the sources.

3.2.2.2 SIMS

SIMS [ACHK93, ArKC96] is a prototype system for data integration from multiple information sources. Instead of performing schema merging of the sources, SIMS adopts an alternative approach, in which a *domain model* of the application domain is defined first, and then each source is described using this model. Notably, in SIMS there is no fixed mapping from a query to the sources; sources are dynamically selected and integrated when the query is submitted. This allows the handling of dynamic information sources, reacting to newly available pieces of information and unexpectedly missing ones. The ability to integrate information sources that are not databases is also pointed out.

The domain model is formalized in terms of a class-based representation language (LOOM). Query processing is performed by using four basic components: (1) a query reformulation component, which identifies the sources required in order to answer the query and the data integration needed; (2) a query access planning component, which builds a plan for retrieving the information requested by the reformulated query; (3) a semantic query-plan optimization component, which both learns the rules for optimizing queries and uses semantic optimization techniques to support multidatabase queries; and (4) an execution component, which executes the optimized query plan.

3.2.2.3 Information Manifold

AT&T's *Information Manifold* (IM) is a prototype system for information gathering from disparate sources such as databases, SGML documents, and unstructured files [LeSK95, KLSS95, LeRO96].

In IM, two components are identified a *world view* and an *information source description* for each source. Both the world view and the information source descriptions are formed essentially by relational schemata. Observe that, although the information sources can be of any kind (not necessarily a relational database), but a view of their data in terms of relations needs to be provided.

The relational schemata of the world view and the information sources are enhanced by sophisticated type descriptions of the attributes of the relations, formulated using the simple description logic of the CLASSIC system [PMB*91]. This allows for forming natural hierarchies among types, reflecting both semantic and structural information. Moreover, various forms of automatic reasoning for query optimization are possible using the inference procedures of description logics. Constraints involving relational schemata and type descriptions of both the world view and the information source descriptions are expressed as Datalog rules in which both relations and type descriptions (interpreted as unary relations) may occur.

In IM, queries are formulated against the world view and answers involve retrieving information from several sources. IM uses Datalog as the query language, enhanced with type descriptions in CLASSIC. Using automatic reasoning on such type descriptions, IM supports optimization techniques for query decomposition that aim at the minimization of the information sources involved in answering a global query, by isolating, for each subquery, those sources that are *relevant*.

It is also possible to specify that an information source has *complete* information on the domain it represents.

Finally, although IM does not implement any specific schema integration strategy, it implicitly enforces one, based on interschema assertions, with a partial global schema. Indeed, IM supports a collection of schemata, one for each source, plus one for the world view. The source schemata are related to the world view by means of constraints which can in fact be seen as interschema assertions. This strategy allows the incremental integration of the information sources and is well suited to deal with their dynamic nature.

3.2.2.4 TSIMMIS

The TSIMMIS project shares with IM the goal of providing tools for the integrated access to multiple and diverse information sources and repositories [CGH*94, Ullm97].

In TSIMMIS, mediators can be conceptually seen as views of data found in one or more sources that are properly integrated and processed. The model used is a simple object-oriented model, the *Object Exchange Model* (OEM). The mediator is defined in terms of a logical language called MSL, which is essentially Datalog extended to support OEM objects. Typically mediators realize virtual views since they do not store data locally. However, it is possible that some mediator materializes the view it provides. Mediators decompose the queries and propagate the resulting subqueries to the wrappers or mediators below them. The answers provided by such levels are then reconstructed through integration steps and processing (using out-of-the-shelf techniques). The TSIMMIS query language is a SQL-like language adapted to deal with OEM objects. Adding a new information source to TSIMMIS requires building of a wrapper for the source and the change of all the mediators that will use the new source. The research within the TSIMMIS project has devised techniques for automatically generating both wrappers and mediators.

It has to be stressed that no global integration is ever performed, in the context of TSIMMIS. Each mediator performs integration independently. As a result, for example, a certain concept may be seen in completely different and even inconsistent ways by different mediators. Such a form of integration can be called *query-based*, since each mediator supports a certain set of queries, for example, those related to the view it serves.

3.2.3 Data Integration – Materialized

The field of data integration with materialized views is the one most closely related to data warehousing. Maintenance of views against updates to the sources is a central aspect in this context, and the effectiveness of maintenance affects the timeliness and the availability of data. In particular, recomputing the view entirely from scratch is often expensive and makes frequent refreshing of views impractical. The study of conditions for reducing overhead in view recomputation is an active research topic. For example, Gupta et al. [GuJM96] introduce the notion of self-maintainability. Self-maintainable views are materialized views which can be updated directly using only log files of the sources.

However, the issues related either to the choice of views to materialize or to the maintenance of materialized views are outside the scope of this chapter, since they pertain more to the physical – rather than the conceptual – part of the design phase of the data warehouse. Here, as in the previous case, we concentrate on aspects related to data integration and proceed by reviewing the current approaches and implementation efforts. The interested reader is referred to Chaps. 4, 6, and 7.

3.2.3.1 Squirrel

The *Squirrel* Project [ZHKF95, ZHKF95a, ZhHK96, HuZh96] provides a framework for data integration based on the notion of *integration mediator*.

In [ZHKF95, ZhHK96], emphasis is placed on data materialization. The Squirrel mediators consist of software components implementing materialized integrated views over multiple sources. A key feature of such components is their ability to incrementally maintain the integrated views by relying on the active capabilities of sources. More precisely, at start-up the mediator informs the source databases with a specification of the incremental update information needed to maintain the views and expects sources to actively provide such information.

Moreover, an automatic generator of Squirrel integrators has been developed. Such a module takes as input a specification of the mediator expressed in a high-level Integration Specification Language (ISL). A mediator specification in ISL includes a description of the relevant subschemata of the source databases and the match criteria between objects of families of corresponding classes, in particular a list of the classes that are matched and a binary matching predicate specifying correspondences between objects of two classes. The output of this module is an implementation of the mediator in the *Heraclitus* language, a database programming language whose main feature is the ability to represent and manipulate collections of updates to the current database state (deltas). In [ZHKF95a], the language *H20* is presented, which is an object-oriented extension of Heraclitus.

Zhou et al. [ZHKF95a] address the problem of object matching in Squirrel mediators. In particular, a framework is presented for supporting intricate object identifier (OID) match criteria, among which, key-based matching, lookup-table-based matching, historical-based-matching, and comparison-based matching are found. The last criterion allows both the consideration of attributes other than keys in object matching and the use of arbitrary Boolean functions in the specification of object matching.

3.2.3.2 WHIPS

WHIPS [HGW*95, WGL*96] is a data warehouse test bed for various integration schemes. The WHIPS architecture consists of a set of independent modules implemented as CORBA objects.

Views are defined by the system administrator in a subset of SQL that includes select-project-join views. The view definition is passed to the *view-specifier* module which parses it into an internal structure, called the *view-tree*, which includes also information from a *metadata store*. The view-tree is sent to the integrator which spawns a view manager that is responsible for managing the view at the data warehouse. The *view manager* initializes the view upon notification by the integrator and computes the changes to the view that become necessary due to updates of the sources. For each source, such updates are detected by a monitor that forwards them to the integrator which, in turn, notifies the relevant *view managers*. The update by the view manager is done by passing appropriate queries to a global *query processor*. The answers are adjusted and combined according to view consistency and maintenance algorithms [ZGHW95] and are then forwarded to the warehouse wrapper.

As already mentioned, the central component of the system is the integrator, whose main role is to facilitate view maintenance. The integrator uses a set of rules which are automatically generated from the view tree, in order to decide which source updates are relevant for which views and therefore have to be forwarded to the corresponding view managers. The current implementation uses a naive strategy which dictates that all modifications to a relation over which a view is defined are relevant to the view. However, this does not take into account, for example, selection conditions, which may render an update irrelevant to a view. Extensions of the integrator in such directions are under development.

3.3 Towards Systematic Methodologies for Source Integration

Many of the research results discussed in the previous sections, such as results on conflict classification, conflict detection, conflict resolution, schema merging, wrapper design, object matching, etc., can be used in a comprehensive solution to the source integration problem. On the other hand, the analysis that we have carried out shows that a general and unified support for incremental source integration in data warehouse with concern on data quality is still missing. As already noticed, the problem of data warehouse construction, and therefore the problem of source integration, is being specifically addressed by the tools for data warehouse schema management and methodologies having their roots in the tradition of database design. However, there seems to be no support for the verification of the validity of interschema assertions, and, more generally, of the specified relationships. We can observe that while the tools are quite powerful in supporting the implementation, there is no comparable support to the design process. The DWQ Project has been concerned with providing a methodological framework for improving the design process of a data warehouse with a particular emphasis on assessing and improving quality factors. We sketch how such a methodological framework supports source integration in data warehousing. We refer the reader to [CDL*97, CDL*98b, CDL*98a, CDL*01] for a detailed treatment.

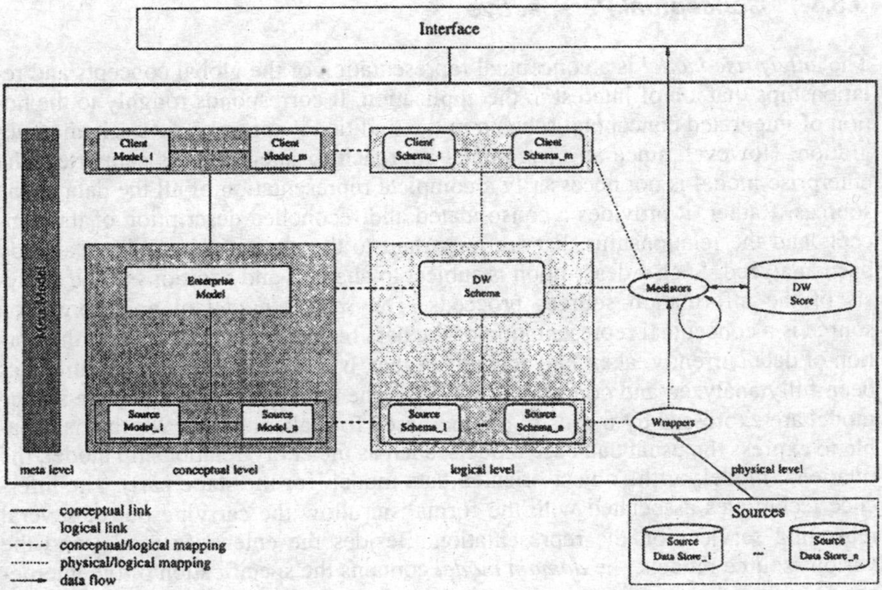

Fig. 3.1. Architecture for source integration

3.3.1 Architecture for Source Integration

The general architecture for source integration in data warehousing is depicted in Fig. 3.1. Three perspectives can be identified in the architecture:

- A *conceptual perspective*, constituted by the *Domain Model* (also called *conceptual data warehouse model*), including an *enterprise model* and one *source model* for each data source, which provides a conceptual representation of both the information sources and the data warehouse;
- A *logical perspective*, constituted by the *Source Schemata* and the *Data Warehouse Schema*, which contains a logical representation of the source and the data warehouse stores, respectively, in terms of a set of definitions of relations, each one expressed through a query over the corresponding conceptual component;
- A *physical perspective*, which consists of the data stores containing the actual data of the sources and of the data warehouse.

In DWQ, source integration exploits automated reasoning techniques to support the incremental building of the conceptual and the logical representations. The designer is provided with information on various aspects, including the global concepts relevant to new information requirements, the sources from which a new view can be defined, the correspondences between sources and/or views, and a trace of the integration steps. The structure of the conceptual and logical perspectives, which constitute the core of the proposed integration framework is outlined below.

3.3.1.1 Conceptual Perspective

The *enterprise model* is a conceptual representation of the global concepts and relationships that are of interest to the application. It corresponds roughly to the notion of integrated conceptual schema in the traditional approaches to schema integration. However, since an incremental approach to integration is supported, the enterprise model is not necessarily a complete representation of all the data of the sources. Rather, it provides a consolidated and reconciled description of the concepts and the relationships that are important to the enterprise, and have already been analyzed. Such a description is subject to changes and additions as the analysis of the information sources proceeds. The *source model* of an information source is a conceptual representation of the data residing in it or at least of the portion of data currently taken into account. Again, it is not required that a source has been fully analyzed and conceptualized. Both the enterprise model and the source model are expressed by means of a logic-based formalism [CDL*98] that is capable to express the usual database models, such as the entity-relationship model, the relational model, or the object-oriented data model (for the static part). The inference techniques associated with the formalism allow the carrying out of several reasoning services on the representation. Besides the enterprise model and the various source models, the *domain model* contains the specification of the interdependencies between elements of different source models and between source models and the enterprise model. The notion of interdependency is central in the architecture. Since the sources are of interest in the overall architecture, integration does not simply mean producing the enterprise model, rather than establishing the correct relationships both between the source models and the enterprise model, and between the various source models. The notion of interdependency is formalized by means of the so called *intermodel assertions* [CaLe93], which provide a simple and effective declarative mechanism to express the dependencies that hold between entities (i.e., classes and relationships) in different models [Hull97]. Again a logic-based formalism is used to express intermodel assertions, and the associated inference techniques provide a means to reason about interdependencies among models.

3.3.1.2 Logical Perspective

Each source, besides being conceptualized, is also described in the *Source Schema* in terms of a logical data model (typically the relational model) which allows the representation of the structure of the stored data. Each object (relational table) of the Source Schema is *mapped* to the conceptual representation of the source (i.e., the source model) by means of a view (i.e., a query) over such a conceptual representation. Objects of the conceptual representation are coded into values of the logical representation. In order to make such coding explicit, additional information, which we call *adornment*, is associated to the relations at the logical level. Such an adornment is also used also in reconciling data coded differently in different sources [CDL*01]. The *Data Warehouse Schema* provides a description of the logical content of the materialized views constituting the data warehouse. The data warehouse schema is typically constituted by a set of relations. Each relation is associated, on the one hand, to a rewriting in terms of relations in the sources, and, on the other hand, to a rewriting in terms of entities at the conceptual level. In both cases adornments are used to facilitate the construction of the rewritings.

3.3.1.3 Mappings

Figure 3.1 explicitly shows the mappings between the conceptual and the logical perspectives, as well as between the logical and the physical perspectives. The mapping between source models and Source Schemata reflects the fact that the correspondence between the logical representation of data in the sources and concepts in the source models should be explicit. The same holds for information needs expressed at the conceptual level and queries expressed at the logical level. Finally, the correspondence between elements of the domain model and the Data Warehouse Schema represents the information about the concepts and the relationships that are materialized in the data warehouse. *Wrappers*, implement the mapping of physical structures to logical structures, and views are actually materialized starting from the data in the sources by means of *mediators* (see Fig. 3.1). The mapping between mediators and query schemata and/or the data warehouse schema explicitly states that each mediator computes the extension of a logical object, which can be either materialized or not. A wrapper is always associated with an element of a source schema, namely, the one whose data are extracted and retrieved by the wrapper. The mapping over the Source Schemata represents exactly the correspondence between a wrapper w and the logical element whose extensional data are extracted from the source through the use of w.

3.3.2 Methodology for Source Integration

We outline a methodology for source integration in data warehousing, based on the three-layered architecture. The methodology deals with two scenarios *source-driven* and *client-driven*.

3.3.2.1 Source-Driven Integration

Source-driven integration is triggered when a new source or a new portion of a source is taken into account for integration. The steps to be accomplished in this case are as follows:

1. *Enterprise and source model construction.* The source model corresponding to the new source is produced, if not available. Analogously, the conceptual model of the enterprise is produced, enriched, or refined.
2. *Source model integration.* The source model is *integrated into the domain model.* This can lead to changes both to the source models, and to the enterprise model. The specification of intermodel assertions and the derivation of implicit relationships by exploiting reasoning techniques, represent the novel part of the methodology. Notably, not only assertions relating elements in one source model with elements in the enterprise model, but also assertions relating elements in different source models are of importance. For example, inferring that the set of instances of a relation in source S_i is always a subset of those in

source S_j can be important in order to infer that accessing source S_i for retrieving instances of the relation is useless.

3. *Source and data warehouse schema specification.* The Source Schema, i.e., the logical view of the new source or a new portion of the source is produced. The relationship between the values at the logical level and the objects at the conceptual level is established. Finally, the mapping between the relations in the schema and the conceptual level is specified by associating each relation to a query over the source model of the source. On the basis of the new source, an analysis is carried out on whether the Data Warehouse Schema should be restructured and/or modified in order to better meet quality requirements. A restructuring of the Data Warehouse Schema may additionally require the design of new mediators.
4. *Data integration and reconciliation.* The problem of data integration and reconciliation arises when data passes from the application-oriented environment to the data warehouse. During the transfer of data, possible inconsistencies and redundancies should be resolved, so that the warehouse is able to provide an integrated and reconciled view of the data of the organization. In our methodology, the problem of data reconciliation is addressed after the phase of data warehouse schema specification and is based on specifying how the relations in the data warehouse schema are linked to the relations in the source schemata. In particular, the methodology aims at producing, for every relation in the data warehouse schema, a specification on how the tuples of such a relation should be constructed from a suitable set of tuples extracted from the relations stored in the sources.

In all the above phases, specific steps for *Quality Analysis* are performed. They are used both to compute the values of suitable quality factors involved in source and data integration and to analyze the quality of the design choices. The quality factors of the conceptual data warehouse model and the various schemata are evaluated and a restructuring of the models and the schemata is accomplished to match the required criteria. This step requires the use of the reasoning techniques associated with our formalisms to check for quality factors such as consistency, redundancy, readability, accessibility, and believability [CDL*97]. Moreover, during the whole design phase, the metadata repository of the data warehouse can be exploited for storing and manipulating the representation of the conceptual model, as well as for querying the metadata in order to retrieve information about the design choices.

3.3.2.2 Client-Driven Integration

The client-driven design strategy refers to the case when a new query (or a set of queries) posed by a client is considered. The reasoning facilities are exploited to analyze and systematically decompose the query and check whether its components are subsumed by the views defined in the various schemata. The analysis is carried out as follows:

1. By exploiting query containment checking, we verify if and how the answer can be computed from the materialized views already stored in the data warehouse.
2. In the case where the materialized information is not sufficient, we test whether the answer can be obtained by materializing new concepts represented in the domain model. In this case, query containment helps to identify the set of subqueries to be issued on the sources and to extend and/or restructure the data warehouse schema. Different choices can be identified, based on various preference criteria (e.g., minimization of the number of sources [LeSK95]) that take into account the above mentioned quality factors.
3. In the case where neither the materialized data nor the concepts in the domain model are sufficient, the necessary data should be searched in new sources, or in new portions of already analyzed sources. The new (portions of the) sources are then added to the domain model using the source-driven approach, and the process of analyzing the query is iterated.

3.4 Concluding Remarks

This chapter has presented the basic process that has to be followed when integrating schemas and data from multiple sources into a data warehouse, a process that – after an initial effort – typically continues throughout the life of a data warehouse. A large number of techniques have been developed to support this process which has also been reviewed. Finally, in the last subsection, we have discussed how these techniques can be extended to cover the conceptual business perspective introduced in Sect. 2.7.

The next two chapters build on these results in two different directions. First, data integration is not a one-shot activity. The data in the sources typically change. This change has to be reflected in the data warehouse, otherwise it will be quickly outdated and useless. This is the subject of Chap. 4. Second, the integrated data format produced for the data warehouse is not necessarily the one according to which analysts want to study the data. Therefore, the question arises how to define multidimensional data models that are better usable for analysts and can still be derived and rederived easily from the basic data warehouse constructed through source integration, data integration, and data refreshment. This topic is treated in Chap. 5.

4 Data Warehouse Refreshment

4.1 What is Data Warehouse Refreshment?

The central problem addressed in this chapter is the refreshment of a data warehouse in order to reflect the changes that have occurred in the sources from which the data warehouse is defined. The possibility of having "fresh data" in a warehouse is a key factor for success in business applications. In many activities, such as in retail, business applications rely on the proper refreshment of their warehouses. For instance, Jahnke [Jahn96] mentions the case of WalMart, the world's most successful retailer. Many of WalMart's large volume suppliers, such as Procter & Gamble, have direct access to the WalMart data warehouse, so they deliver goods to specific stores as needed. WalMart pays such companies for their products only when they are sold. Procter & Gamble ships 40% of its items in this way, eliminating paperwork and sale calls on both sides. It is essential for the supplier to use fresh data in order to establish accurate shipment plans and to know how much money is due from the retailer. Another example is Casino Supermarche, in France, which recouped several millions dollars when they noticed that Coca-Cola was often out of stock in many of their stores. Freshness of data does not necessarily refer to the highest currency but the currency required by the users. Clearly, applications have different requirements with respect to the freshness of data.

4.1.1 Refreshment Process within the Data Warehouse Lifecycle

The data warehouse can be defined as a hierarchy of data stores which goes from source data to highly aggregated data (often called data marts). Between these two extremes can be other data stores depending on the requirements of OLAP applications. One of these stores is the *Corporate Data Warehouse* store (CDW) which groups all aggregated views used for the generation of the data marts. The corporate data store can be complemented by an Operational Data Store (ODS) which groups the base data collected and integrated from the sources. Data extracted from each source can also be stored in different data structures. This hierarchy of data stores is a logical way to represent the data flow between the sources and the data marts. In practice, all the intermediate states between the sources and the data marts can be represented in the same database.

We distinguish four levels in the construction of the hierarchy of stores. The first level includes three major steps: (a) the extraction of data from the operational data sources, (b) their cleaning with respect to the common rules defined for the data warehouse store, and (c) their possible archiving in the case when integration needs some synchronization between extractors. Note, however, that this decomposition is only logical. The extraction step and part of the cleaning step can be grouped into the same software component, such as a wrapper or a data migration tool. When the extraction and cleaning steps are separated, data need to be stored in between. This can be done using one storage medium per source or one shared medium for all sources.

The second level is the integration step. This phase is often coupled with rich data transformation capabilities into the same software component, which usually performs the loading into the ODS when it exists or into the CDW. The third level concerns the data aggregation for the purpose of cubes construction. Finally, the fourth level is a step of cube customization. All these steps can also be grouped into the same software, such as a multidatabase system. A typical operational view of these components is portrayed in Fig. 4.1. All data that are input to the integration component use the same data representation model.

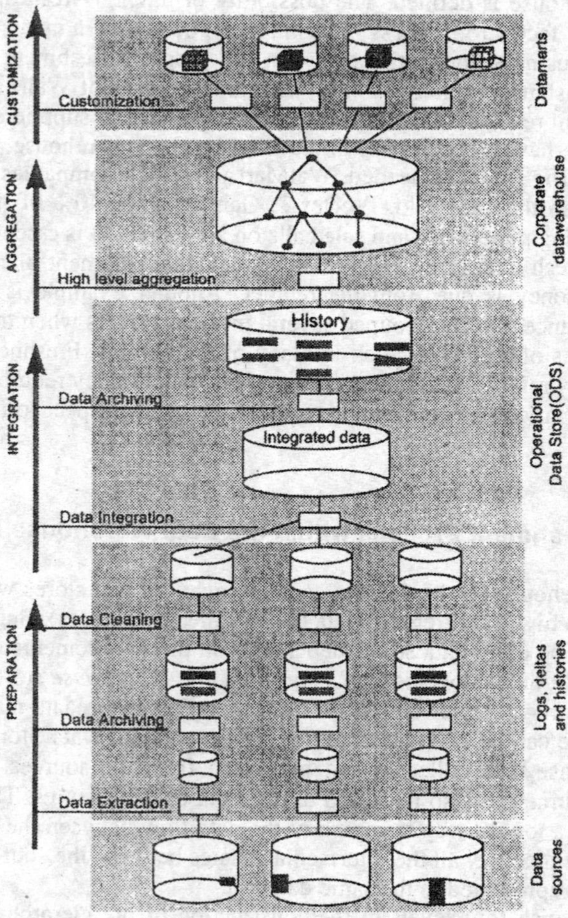

Fig. 4.1. Operational view of the construction of a data warehouse

In order to understand which kind of tools the refreshment process needs, it is important to locate it within the global data warehouse lifecycle which is defined by the three following phases:

- The *design phase* consists of the definition of user views, auxiliary views, source extractors, data cleaners, data integrators, and all others features that guarantee an explicit specification of the data warehouse application. As suggested in Chap. 7, these specifications could be done with respect to abstraction levels (conceptual, logical, and physical) and user perspectives (source view, enterprise view, client views). The result of the design is a set of formal or semiformal specifications which constitute the metadata used by the data warehouse system and applications.
- The *loading phase* consists of the initial data warehouse instantiation, which is the initial computation of the data warehouse content. This initial loading is globally a sequential process of four steps: (1) preparation, (2) integration, (3) high level aggregation, and (4) customization. The first step is done for each source and consists of data extraction, data cleaning, and possibly data archiving before or after cleaning. The second step consists of data integration, which is reconciliation of data originated from heterogeneous sources and derivation of the base relations (or base views) of the ODS. The third step consists of the computation of aggregated views from base views. In all three steps not just the loading of data but also the loading of indexes is of crucial importance for query and update performance. While the data extracted from the sources and integrated in the ODS are considered as ground data with very low-level aggregation, the data in aggregated views are generally highly summarized using aggregation functions. These aggregated views constitute what is sometimes called the CDS, i.e., the set of materialized views from which data marts are derived. The fourth step consists of the derivation and customization of the user views which define the data marts. Customization refers to various presentations needed by the users for multidimensional data. Figure 4.2 shows the flow of data within the sequential process of loading. This is a logical decomposition whose operational implementation receives many different answers in the data warehouse products.
- The *refreshment phase* has a data flow similar to the loading phase but, while the loading process is a massive feeding of the data warehouse, the refreshment process captures the differential changes that occurred in the sources and propagates them through the hierarchy of data stores. The preparation step extracts from each source the data that characterize the changes that have occurred in this source since the last extraction. As for the loading phase, these data are cleaned and possibly archived before their integration. The integration step reconciles the source changes coming from multiple sources and adds them to the ODS. The aggregation step recomputes incrementally the hierarchy of aggregated views using these changes. The customization step propagates the summarized data to the data marts. As previously mentioned, this is a logical decomposition whose operational implementation receives many different answers in the data warehouse products. This logical view allows a certain traceability of the refreshment process.

The difference between the refreshment phase and the loading phase is mainly in the following. First, the refreshment process may have a complete asynchronism between its different activities (preparation, integration, aggregation, and customization). Second, there may be a high-level parallelism within the preparation activity itself, with each data source having its own window of availability and strategy of extraction. The synchronization is done by the integration activity. Another difference lies in the source availability. While the loading phase requires a long period of availability, the refreshment phase should not overload the operational applications that use the data sources. Then, each source provides a specific access frequency and a restricted availability duration. Finally, there are more constraints on response time for the refreshment process than for the loading process. Indeed, with respect to the users, the data warehouse does not exist before the initial loading, so the response time is confused with the project duration. After the initial loading, data become visible and should satisfy user requirements in terms of data availability, accessibility and freshness.

4.1.2 Requirements and Difficulties of Data Warehouse Refreshment

The refreshment of a data warehouse is an important process which determines the effective usability of the data collected and aggregated from the sources. Indeed, the quality of data provided to the decision makers depends on the capability of the data warehouse system to propagate the changes made at the data sources in reasonable time. Most of the design decisions are then influenced by the choice of data structures and updating techniques that optimize the refreshment of the data warehouse.

Building an efficient refreshment strategy depends on various parameters related to the following:

- *Application requirements* (e.g., data freshness, computation time of queries and views, data accuracy)
- *Source constraints* (e.g., availability windows, frequency of change)
- *Data warehouse system limits* (e.g., storage space limit, functional limits)

Most of these parameters may evolve during the data warehouse lifetime, hence leading to frequent reconfiguration of the data warehouse architecture and changes in the refreshment strategies. Consequently, data warehouse administrators must be provided with powerful tools that enable them to efficiently redesign data warehouse applications.

For those corporations in which an ODS makes sense, Inmon [Inmo96] proposes to distinguish among three classes of ODSs, depending on the speed of refreshment demanded.

- The first class of ODSs is refreshed within a few seconds after the operational data sources are updated. Very little transformations are performed as the data passes from the operational environment into the ODS. A typical example of such an ODS is given by a banking environment where data sources keep individual accounts of a large multinational customer, and the ODS stores the total balance for this customer.

- With the second class of ODSs, integrated and transformed data are first accumulated and stored into an intermediate data store, and then periodically forwarded to the ODS on, say, an hourly basis. This class usually involves more integration and transformation processing. To illustrate this, consider now a bank that stores in the ODS an integrated individual bank account on a weekly basis, including the number of transactions during the week, the starting and ending balances, the largest and smallest transactions, etc. The daily transactions processed at the operational level are stored and forwarded on an hourly basis. Each change received by the ODS triggers the updating of a composite record of the bank account throughout the current week.
- Finally, the third class of ODSs is strongly asynchronous. Data are extracted from the sources and used to refresh the ODS on a day-or-more basis. As an example of this class, consider an ODS that stores composite customer records computed from different sources. As customer data change very slowly, it is reasonable to refresh the ODS in a more infrequent fashion.

Quite similar distinctions also apply for the refreshment of a global data warehouse, except that there is usually no counterpart for ODS of the first class. The period for refreshment is considered to be larger for global data warehouses. Nevertheless, different data warehouses demand different speeds of refreshment. Besides the speed of the refreshment, which can be determined statically after analyzing the requirements of the information processing application, other dynamic parameters may influence the refreshment strategy of the data warehouse. For instance, one may consider the volume of changes in the data sources, as given by the number of update transactions. Coming back to the previous example of an ODS of the second class, such a parameter may determine dynamically the moment at which the changes accumulated into an intermediate data store should be forwarded to the ODS. Another parameter can be determined by the profile of queries that execute on the data warehouse. Some strategic queries that require to use fresh data may entail the refreshment of the data warehouse, for instance using the changes that have been previously logged between the sources and the ODS or the sources and the global data warehouse.

In any case, the refreshment of a data warehouse is considered to be a difficult and critical problem for three main reasons.

- First, the volume of data stored in a warehouse is usually large and is predicted to grow in the near future. Recent inquiries show that 100 GB warehouses are becoming commonplace. Also, a study from META Group published in January 1996 reported that 52% of the warehouses surveyed would be 20 GB to 1 TB or larger in 12-18 months. In particular, the level of detail required by the business leads to fundamentally new volumes of warehoused data. Further, the refreshment process must be propagated along the various levels of data (ODS, CDW, and data marts), which enlarges the volume of data that must be refreshed.
- Second, the refreshment of warehouses requires the execution of transactional workloads of varying complexity. In fact, the refreshment of warehouses yields different performance challenges depending on its level in the architecture. The refreshment of an ODS involves many transactions that need to access and update a few records. This is best illustrated by previous examples of ODSs that keep composite records. Thus, the performance requirements for refreshment

are those of general-purpose record-level update processing. The refreshment of a global data warehouse involves heavy load and access transactions. Possibly large volumes of data are periodically loaded in the data warehouse, and once loaded, these data are accessed either for informational processing or for refreshing the local warehouses. Power for loading is now measured in GB per hour, and several companies are moving to parallel architectures, when possible, to increase their processing power for loading and refreshment. The network interconnecting the data sources to the warehouse can also be a bottleneck during refreshment and calls for compression techniques for data transmission. Finally, as a third reason, the refreshment of local warehouses involves transactions that access many data, perform complex calculations to produce highly summarized and aggregated data, and update a few records in the local warehouses. This is particularly true for the local data warehouses that usually contain the data cubes manipulated by OLAP applications. Thus, a considerable processing time may be needed to refresh the warehouses. This is a problem because there is always a limited time frame during which the refreshment is expected to happen. Even if this time frame goes up to several hours and does not occur at peak periods (say, at night), it may be challenging to guarantee that the data warehouse will be refreshed within it.

- Third, the refreshment of a warehouse may be run concurrently with the processing of queries. This may happen because the time frame during which the data warehouse is not queried is either too short or nonexistent (e.g., when the data warehouse is accessed by users located in different hemispheres within worldwide organizations). As noted by Red Brick [RBSI96], the batch windows for loading and refreshment shrink as system availability demands increase. Another argument is the need to run decision-support queries against fresh data, as showed by the earlier examples in retail. Thus, the problem is to refresh the data warehouse without impeding the traffic of data warehouse queries too much. A priori, the two processes, refreshing and querying, conflict because refreshing writes to the data warehouse while querying reads the same data. An analysis of the performance problems with ODS workloads is carried out in [Inmo96].

In summary, the refreshment of data warehouses is an important problem because it directly impacts the quality of service offered to integrated operational processing or informational processing applications. It is a difficult problem because it entails critical performance requirements, which are hard to achieve. Quoting Inmon, "Speed of refreshment costs, and it costs a lot, especially in the world of operational data stores." It is difficult to engineer refreshment solutions because the requirements which they must comply with may vary over the time or can be subject to dynamic parameters.

4.1.3 Data Warehouse Refreshment: Problem Statement

Our analysis of the data warehouse refreshment problem follows the same distinction of levels as we used for data warehouse loading. Indeed, we view the refreshment problem as an incremental data warehouse construction process. Figure 4.2 gives an operational view of the refreshment process. Incrementality occurs at various steps. First, the extraction component must be able to output and record the changes that have occurred in a source. This raises issues such as:

- The detection of changes in the data sources
- The computation and extraction of these changes
- The recording of the changes

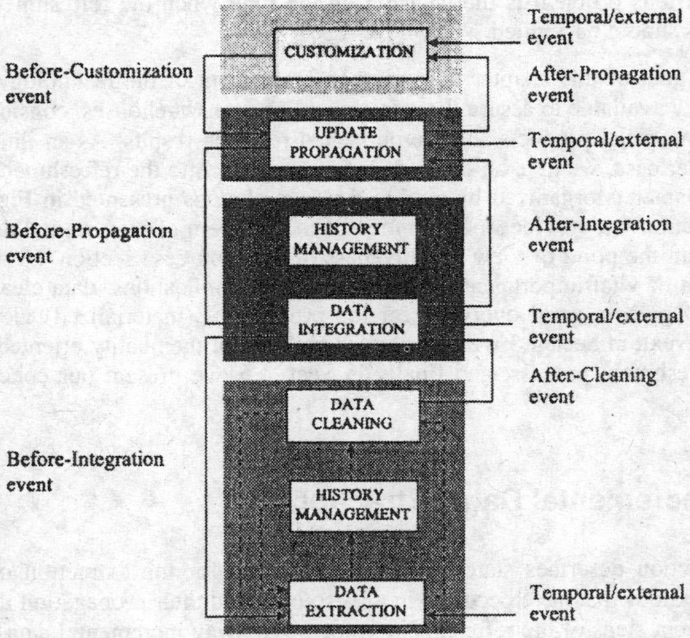

Fig. 4.2. The workflow of the refreshment process

The answer to these issues clearly depends on the functionality of a source and its availability. From a performance perspective, it is critical to isolate the modified data from the sources as early in the extraction process as possible. This will drastically reduce the amount of data to be migrated towards the data warehouse or ODS. Second, integration must be incremental. Data transformations must be completed incrementally (e.g., data cleaning). A more difficult problem is to generate the operations that must be applied to the intermediate data stores or to the ODS. Knowing the data that have changed in a source, the following several problems have to be tackled:

- The computation of the data that must be changed in the warehouse,
- The estimation of the volume of information needed from the other sources to compute the new value of the data warehouse,
- The estimation of the time needed for this computation, and
- The estimation of the time needed for the actual update of the data warehouse.

Finally, the last problem is the incremental loading of data in order to reduce the volume of data that has to be incorporated into the warehouse. Only the updated or inserted tuples are loaded. This raises several issues:

- The incremental loading transaction may conflict with queries and may have to be chopped into smaller transactions.
- Refreshment transactions must be synchronized so that the views accessed by queries correspond to a consistent snapshot.
- A serious problem is the decision on the time when the refreshment transactions should be applied.

The goal of this chapter is to give an assessment of the technologies that are currently available to assure the refreshment of data warehouses, considering both the commercial products and the published research results. As an illustration of this latter case, we give an overview of our approach to the refreshment problem. This chapter is organized by tracing the several steps presented in Figs. 4.1 and 4.2. Section 4.2 provides a description of the problem of incremental data extraction from the point of view of activeness of data sources. Section 4.3 discusses a problem of vital importance for data warehouse applications: data cleaning. Section 4.4 gives a good overview for the problem of (materialized) view maintenance. Next, in Sect. 4.5 we describe the problem of the quality oriented design of the refreshment process, and finally in Sect. 4.6, we present our concluding remarks.

4.2 Incremental Data Extraction

This section describes state-of-the-art techniques for the extraction of relevant modifications that have occurred in data sources and their propagation them to the subsequent steps of the refreshment process. The way incremental data extraction can be implemented depends on the characteristics of the data sources and also on the desired functionality of the data warehouse system.

Data sources are heterogeneous and can include conventional database systems and nontraditional sources like flat files, XML and HTML documents, knowledge systems, and legacy systems. The mechanisms offered by each data source to help the detection of changes are also quite heterogeneous.

Following existing work on heterogeneous databases [CGH*94, TAB*97], it is convenient to associate a wrapper with every data source in order to provide a uniform description of the capabilities of the data sources. Moreover, the role of the wrapper in a data warehouse context is enlarged. Its first functionality is to give a description of the data stored by each data source in a common data model. In the rest of this section, we assume that this common model is a relational data model. This is the typical functionality of a wrapper in a classical wrapper/mediator architecture; therefore, we shall call it *wrapper functionality*. The second functionality is to detect (or extract) the changes of interest that have happened in the underlying data source. This is a specific functionality required by data warehouse architectures in order to support the refreshment of the data warehouse in a incremental way. For this reason, we reserve the term *change monitoring* to refer to this kind of functionality.

4.2.1 Wrapper Functionality

The principal function of the wrapper, relative to this functionality, is to make the underlying data source appear as having the same data format and model that are used in the data warehouse system. For instance, if the data source is a set of XML documents and the data model used in the data warehouse is the relational model, then the wrapper must be defined in such a way so that it presents the data sources of this type as if they were relational.

Recently, the development of wrapper generators has received attention from the research community, especially in the case of sources that contain semi-structured data such as HTML or SGML documents. These tools, for instance, enable to query the documents using an OQL-based interface. In [HBG*97], a wrapper implementation toolkit for quickly building of wrappers is described. This toolkit tries to minimize the work of the wrapper implementor just to the construction of a few specialized components in a preconstructed wrapper architecture. The work involved in those few specialized components depends on the type of the data source.

Another important function that should be implemented by the wrapper is to establish the communication with the underlying data source and allow the transfer of information between the data source and the change monitor component. If the data warehouse system and the data source share the same data model, then the function of the wrapper would be just to translate the data format (if different) and to support the communication with the data source. For data sources that are relational systems, and supposing that the data model used in the data warehouse is also relational, it is possible to use wrappers that have been developed by software companies, such as database vendors or database independent companies. These wrappers, also called "middleware," "gateways," or "brokers," have varying capabilities in terms of application programming interface, performance, and extensibility.

In the client-server database environment, several kinds of middleware have already been developed to enable the exchange of queries and their associated answers between a client application and a database server, or between database servers, in a transparent way. The term "transparent" usually means that the middleware hides the underlying network protocol, the database systems, and the database query languages supported by these database systems from the application.

The usual sequence of steps during the interaction of a client application and a database server, through a middleware agent is as follows. First, the middleware enables the application to connect and disconnect to the database server. Then, it allows the preparation and execution of requests. A request preparation specifies the request with formal parameters, which generally entails its compilation in the server. A prepared request can then be executed by invoking its name and passing its actual parameters. Requests are generally expressed in SQL. Another functionality offered by middleware is the fetching of results, which enables a client application to get back all or part of the result of a request. When the results are large, they can be cached on the server. The transfer of requests and results is often built on a protocol supporting remote procedure calls.

There has been an important effort to standardize the programming interface offered by middleware and the underlying communication protocol. Call Level Interface (CLI) is a standardized API developed by the X/Open standardization committee [X/Op92]. It enables a client application to extract data from a relational database server through a standard SQL-based interface. This API is currently supported by several middleware products such as ODBC [MiPE94], and IDAPI (Borland). The RDA standard communication protocol specifies the messages to be exchanged between clients and servers. Its specialization to SQL requests enables the transport of requests generated by a CLI interface.

Despite these efforts, existing middleware products do not actually offer a standard interface for client-server developments. Some products such as DAL/DAM (Apple) or SequeLink (Techgnosis) offer their own API, although some compatibility is sometimes offered with other tools, such as ODBC. Furthermore, database vendors have developed their own middleware. For instance, Oracle proposes several levels of interface, such as Oracle Common Interface (OCI), on top of its client-server protocol named SQL*Net. The OCI offers a set of functions close to the ones of CLI, and enables any client having SQL*Net to connect to an Oracle server using any kind of communication protocol. Although an ODBC interface can be developed on top of OCI, this solution offers poor performance as compared to the use of the database vendor middleware. Oracle also provides "transparent gateways" for communicating between an Oracle server and other kinds of database servers (e.g., IMS, DB2) and "procedural gateways" to enable a PL/SQL program to invoke external programs using RPCs. Other database vendors have developed similar strategies. For instance, Sybase offers a library of functions, called DB-Library, for client-server communications, and an interface, called Open Data Services (ODS), for server-to-server communications. Using ODS, several gateways to other servers have been developed.

Finally, an alternative way to provide a transparent access to database servers is to use Internet protocols. In fact, it must be noted that the World Wide Web is simply a standards-based client-server architecture. It holds major advantages over the above client-server application environments in its ability to integrate diverse clients and servers without additional development and at a lower cost of implementation. Therefore, Internet and intranets arise as a promising standards-based technology for the extraction of data from multiple sources.

4.2.2 Change Monitoring

The detection of changes in the sources is implemented by a specialized component in the wrapper which we will call *change monitor*. The responsibility of the change monitor is to detect changes of interest that occurred in the underlying data source and propagate them to other modules of the data warehouse system also involved in the refreshment process. We start this section by introducing some useful definitions; next, we present techniques that can be used for detect relevant changes of data in data sources.

Let *S* be the set of all data instances stored by a source, later called a *source state*, and *O* a sequence of data modification operations. *O(S)* denotes the new source state resulting from the application of the sequence *O* to the initial source state *S*. Moreover, if these operations in *O* are durable, then we shall say that *O(S)* is an *observable state*. By durable operation, we mean an operation that will be eventually applied to the source even in the presence of failures, such as memory or disk failure. We use the term durable by analogy to durable database transactions. A *change* monitored by the change monitor is the difference between two consecutive observable source states. This means that the only changes observable by the change monitor are durable changes.

Following the work in [Wido95, KoLo98], it is possible to classify the data sources according to the support (mechanisms) that they provide for helping to solve the change detection problem. Evidently, the more support is provided; the simpler is the implementation of the change monitor. Figure 4.3 shows this data source classification. There are two main groups of data source: *cooperative* and *non-cooperative* data sources. By cooperative data sources we mean data sources that supply a mechanism that allows the automatic detection of changes in the data source and the respective notification, like triggers or ECA-rules in the case of *active* data sources.

A data source may offer more than one mechanism to support the detection of changes in the data source. For example, the Oracle relational database system can be seen as an active data source, a queryable data source or a logged data source. The choice of the right mechanism will depend on the availability constraints of the data source and desired characteristics of the data warehouse system, specially the refreshment interval.

The methods used to compute the changes in a source fall into two categories: *external* and *intrusive* methods. An external method computes the changes without using any particular source functionality while intrusive methods try to exploit the capabilities of the source. Almost all methods described here fall into the intrusive category; the only method that can be classified as external is the one that use the snapshots of a data source to compute its changes of interest.

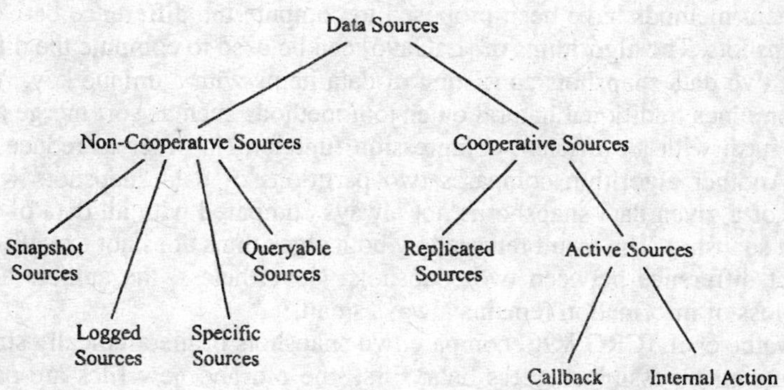

Fig. 4.3. Data source classification

Two parameters influence the design of the incremental extraction strategy. On the one hand, we can consider the technical features offered by a legacy system, or a source in general, for trapping the changes that have occurred. On the other hand, we find the availability of data and metadata (i.e., the schema) of the data source to the change monitor. For instance, the change monitor may only be allowed to access the data source in a read-only mode, without having the right to modify it. Another example is that the change monitor can only access the data source at certain points in time, like only at night.

The C2offein system [KoLo98] allows the definition of ECA-rules for complex situations in a distributed and heterogeneous CORBA-based system. For each data source involved in the system, a wrapper (that can be seen as our change monitor) is defined, making the source appear as an active source. The system uses a source classification similar to the one given above and provides an architecture for wrapper generation where the developer only has to write the code that specifies the particularities of the data source. The amount of code that must be specified by the developer depends on type of the underlying data source.

4.2.2.1 Snapshot Sources

Snapshot sources are sources that provide their content in bulk, i.e., without any selective capability. The only way to access to the data source content is through a dump of its data. A typical example of this type of data source is ordinary files.

The method used by the change monitor to detect the changes is basically the following: the change monitor periodically reads the current state (snapshot) of the source, compares it with the previous state of the data source using some specific algorithm that extracts the occurred changes, and then it chooses just the relevant changes. The change monitor must keep a version of the snapshot, corresponding to the previous state of the snapshot, which should be updated after the comparison has been done. This strategy induces a minor overhead on the data source, but the amount of data transmitted may be very large. Moreover, the computation of the changes may be very expensive, depending principally of the size of the snapshot. This method is not very scalable with respect to the size of the data source.

Efficient methods have been proposed to compute the difference between two data snapshots. The algorithms of [LaGa96] can be used to compute the difference between two data snapshots consisting of data items with a unique key. An algorithm combines traditional natural outer-join methods such as sort merge and partitioned hash with an efficient compression function (in order to reduce the I/O costs). Another algorithm compares two partitions of data snapshots where an element of a given data snapshot is not always compared with all data of the second data snapshot. The result returned by both algorithms does not exactly provide the exact difference between two snapshots. Nevertheless, the authors advocate that the loss of information remains always small.

Chawathe et al. [CRGW96] compare two snapshots of hierarchically structured files containing possibly keyless data. First, the old and new files are parsed to produce trees. Each node in a tree has a value and a label. The value represents a piece of text in the file, and the label gives the hierarchical position of the corresponding data. Then, the trees are compared with respect to the labels. The comparison algorithm determines whether two pairwise compared nodes are considered to be "similar." It uses metrics specified by a pair ("acceptability," "sufficiency"). The acceptability gives the amount of differences between the values of the nodes, and the sufficiency gives the amount of differences between the structures of the sub-trees rooted at the nodes. Two nodes are considered as "similar" if both the acceptability and sufficiency do not exceed a threshold given as input of the algorithm. Finally, given two trees T_1 and T_2 where some nodes are not similar, the algorithm computes the cheapest set of operations O such that $O(T_1)$ is similar to $O(T_2)$.

4.2.2.2 Specific Sources

In this type of data source, each data source is a particular case. Consequently, it is not possible to use a general method. Legacy systems are the most prominent example of this type of data sources.

The detection method must make use of specific data existing in the source. For instance, some legacy systems are able to write *delta files* that describe their actions, for the purpose of auditing. It is then possible to use these delta files to detect the interesting changes. In some ideal cases, the legacy system timestamps all of its underlying records and is capable of isolating and providing the changed records in one simple extraction step.

4.2.2.3 Logged Sources

Logged sources have a log file where all their actions are registered. Typical examples of logged sources are database systems or mail systems.

For logged sources, the changes of interest will be detected by periodic polling the log of the data source which must then be further analyzed. However, the log is not easily accessible (system administration privileges are required) and records are hard to interpret (no standard format). Moreover, if the log to be analyzed is the single log file used by a database system for recovery, then the detection of the relevant changes will be a slow process because every change that happened in the data source, independently of being of interest or not, will be registered on the log. Also, as the log file size may be big, there may be an overhead in its transmission. Nevertheless, some logged sources allow the specification of log files for specific data items. Therefore, there will be a reduction in the time overhead for the detection of the relevant changes but at the same time there will be an overhead in space for storing these new specific log files.

4.2.2.4 Queryable Sources

A queryable source is a data source that offers a query interface. For this type of data source, the detection of relevant changes is done by periodic polling of the interesting data items of the data source and by comparing them with the previous version. Examples of queryable sources are relational database systems.

The detection method in this case consists in polling the data source periodically for each relation of interest. At each polling, the new version of the relation that must be compared with the previous version is obtained. The comparison is based on the keys of the relation. For instance, an updated tuple is identified if there is a tuple in both versions of the relation with identical keys but with at least a different field in the remainder fields of the relation. Some changes may not be detected by this method. For example, if an insertion of a tuple and its deletion occur during a polling interval then both changes are not detected. This method only detects the net-effect of the changes that have occurred since the last polling.

There is also a performance issue related with the polling frequency. If the frequency is too high then the performance will degrade, but if the frequency is too low then changes of interest may not be detected in a timely way.

4.2.2.5 Replicated Sources

Replicated sources are data sources that already supply a replication service or allow the use of a replication system. The detection of interesting changes is done by analyzing the messages sent by the replication system that reflect the changes occurred in the data source.

Each copy of the database is called a *resource replica*, or *replica*, for short, in this chapter. In this case, the master replica, also called the *primary* copy, is held by the source and we can say that a *secondary* copy is held by the change monitor. Updates only occur at the source, which sends them to the secondary copy in the same order in which they have been executed. By processing these updates in the same order, the change monitor is able to detect the relevant changes that have occurred in the data source.

Replication systems which offer sophisticated mechanisms to synchronize updates occurring from primary and secondary copies are not necessarily fast and are not useful in our case because the updates only happen in the primary copy. It is better to use an asynchronous replication scheme where the updates to the primary copy generate a stream of updates to the secondary copy. These updates are only propagated to the secondary copy after the update transaction on the source has committed. As another point, most of the replication systems are not open and extensible. For instance, their update propagation mechanism cannot be finely parameterized, so as to tune which data items should be replicated. This situation resembles the case of logged sources, where the change monitor has to analyze all the occurred changes in order to choose the interesting subset of them. Finally, few replication systems are able to manage arbitrary sources [e.g., Data Propagator (IBM), OpenIngres 2.0 (Computer Associates], while others are limited to handling only relational databases.

To propagate the updates to the secondary copy, the replication system may push the stream of updates towards the secondary copy by calling remote procedures at the site of the secondary copy, which in our case is the change monitor. This solution is implemented by systems like Informix, Ingres, Oracle, or Sybase. Another solution, e.g., implemented by IBM Data Propagator, is that the change monitor pulls the stream of updates using a specific query. The update propagation may happen manually, periodically, or using specific criteria (e.g., when a certain amount of data has been modified in a primary copy).

4.2.2.6 Callback Sources

Callback sources are sources that provide triggers or other active capabilities [WiCe95] and so they are able to automatically detect changes of interest and notify those changes to the interested parties. In this particular case, the interested parties are entities outside the data source.

For each interesting data item, one or more triggers must be defined in the data source so that when a data item is modified, the trigger is fired and the change monitor is notified automatically of this change. For active relational database systems, we have to define three triggers, for each interesting relation: one concerning insertions in the relation, another concerning updates, and a third concerning delete operations over the relation.

Oracle7 relational database system is an example of a callback source. This system allows the communication of Oracle sessions with our change monitor using Oracle Pipes [OrCo96]. Pipes are an input/output (I/O) mechanism where messages are written and read in a FIFO order. At the beginning of the detection process a pipe is created between the Oracle system and the change monitor. The Oracle sessions will be the writer in this pipe and the change monitor will be the reader. The triggers are programmed so that when they are fired, their action will write a message in the pipe that will describe exactly the occurred modification. The change monitor (or a thread in the change monitor) is always trying to read messages from the pipe, so each message written in the pipe is read and processed immediately by the change monitor. When there is no message in the pipe, the change monitor is blocked.

This detection method has the advantage of detecting the interesting changes without delay and the involved overhead is negligible. One disadvantage is the lack of standardization for the callback mechanism. Another disadvantage is that this method modifies the data source (by defining the necessary triggers), which may not be always feasible.

4.2.2.7 Internal Action Sources

Internal action sources are similar to callback sources, except that it is not possible to define a trigger whose action may have external database effects.

This method requires both to define triggers in the data source and to create auxiliary relations, henceforth called *delta tables*. To illustrate the principles of this method, consider a relation *R* in the data source (we are considering an active relational database system), whose modification we are interested in. First, a new relation, say *deltaR,* must be created in the data source. This relation will have at least the same fields as relation *R*. It can have new fields that will allow supplying more information about the modification, as the time when the modification occurred, or the user that issued the modification (this can also be made for callback sources). Then, a trigger is defined on *R* for each type of data modification (insert, delete, and update) that may occur. The purpose of each trigger is to record the changes performed by the corresponding data modification statement into *deltaR*. The change monitor can then periodically read the delta tables and compute the changes accordingly. Since the triggers are executed within the update transaction that caused their execution, the change monitor always reads a change in the sense of our earlier definition.

A basic algorithm would work as follows. The change monitor keeps the last accessed state of the delta relations. Then, periodically, it reads the current state of the delta relations, and computes the difference between the current and the old state in order to derive the new modifications that will be propagated to the data warehouse system. An improved version of this algorithm may use a counter to identify the new tuples in the delta relations. Suppose that an extra field with the counter value is added to each delta relation. The action of each trigger is modified so that the trigger first accesses the counter to get its current value, and then uses it to create the tuples to insert into the delta table. The counter is automatically incremented each time it is accessed. The change monitor stores the maximum value of the counter, say *oldC*, read from the last time it queried the delta relations. Then periodically, the change monitor reads all tuples from the delta relations that have a counter value strictly greater than *oldC*. Finally, a change is computed and *oldC* is updated. Since each tuple of the delta relation is read only once, the change monitor may invoke a stored procedure in the source DBMS for purging the delta tables at predefined time instants (e.g., at night). The optimized algorithm improves the basic one in several ways: it decreases the amount of data that needs to be read by the change monitor, it avoids storing an old snapshot of the delta relations in the change monitor, it minimizes the volume of the delta relations at the source, and it decreases the cost of computing the changes. Nevertheless, as in the case of the queryable sources, there is always the problem of choosing the right polling frequency. If the frequency is too high the performance will degrade, whereas, if it is too low the data warehouse may not be refreshed in time. Another drawback is that the data source may refuse the intrusion at the level of its schema, which happens quite often in practice with legacy applications [SiKo95].

4.3 Data Cleaning

Data Cleaning (also called *Data Cleansing* or *Data Scrubbing*) is fundamental in almost all data migration processes. Its main concern is the quality of data obtained at the end of the migration. Data cleaning is particularly important when data comes from heterogeneous data sources that may not share the same data schema or may represent the same real entity in different ways.

The cleaning process constitutes part of the data transformation stage in the construction of a data warehouse. In general, data transformation defines how the data residing in operational systems are converted into data in the warehouse reflecting the business needs. In particular, besides data cleaning, data transformation involves the integration of different values and formats according to transformation tables, the validation of integrated data according to the warehouse schema, and data summarization and aggregation according to the application requirements. It is based on a set of metadata that provides structural information for building the warehouse such as the mapping between source and target schemata and data encoding formats. Chap. 3 describes in detail the schema and data integration step. This chapter also mentions the case of data reconciliation which is some kind of data cleaning at the data integration level.

In order to support business decisions well, data stored in a data warehouse must be accurate, relevant to the business requirements, consistent across redundant sources, and not lacking in information [Will97]. Hence, data migrated from distinct data sources must be treated so that the level of data quality obtained is the maximum possible according to the demands of the data warehouse applications. Usually, before being transformed, data is said to be "dirty." Dirtiness of data can be verified on a per source basis such as a mismatch between the source data format and the expected integrated data format for the same field, or on a multisource basis as it happens when merging information from two sources about the same entity and this information is inconsistent. From now on, we refer to cases of dirty data without discriminating between per source or multisource basis. In general, the most common examples of dirty data [Hurw97] are as follows:

- Different formats of data for the same field (for instance the information about the state in a location field can be the state abbreviation, the state name, or even the state code);
- Free-form text may hide some important information as the "C/O" specification that may appear in a name or an address field;
- Inconsistent values for the same entity due to typing errors;
- Mismatch between values and the corresponding field description (for example a name field may contain commercial as well as personal names);
- Missing values that must be filled in according to the warehouse schema; and
- Duplicated information that may arise within the same source as well as when two sources provide exactly the same information about the same entity but using a different key.

Existing data warehouse and data migration tools attempt to solve such problems through their cleaning modules which offer the following functionalities:

- *Conversion and normalization functions* that transform and standardize heterogeneous data formats
- *Special-purpose cleaning* that cleans specific fields using dictionaries to look for synonyms
- *Domain-independent cleaning* that applies field matching algorithms to equivalent fields from different sources in order to decide on their matching
- *Rule-based cleaning* that is based on a set of so-called "business rules" that specify the conditions on which two values from different sources match

The last two cleaning methods apply to the case where the integrated data resides in different sources that have to be merged in order to populate the data warehouse. Data format standardization and field cleaning can be performed on a per-source or on a multisource basis. In the next sections, we will describe these functionalities and we will indicate some of the commercial data warehouse tools that implement them. For further details, the reader is referred to recent overviews of research in data cleaning [RaHa00] and data transformations [Rund99].

4.3.1 Conversion and Normalization Functions

Given a common data format in the integrated view of the data warehouse, most of the warehouse tools provide a module that converts distinct data formats into the expected one. The simplest example is the SQL*Loader module of Oracle [OrCo96] that transforms data from external files or tables into and Oracle database. In brief, SQL*Loader loads data in several formats, filters them and loads the result in tables. Another way of converting formats consists in attaching a software module called *wrapper* or *translator* to each source in order to export information to the data warehouse. The wrapper provides an interface to translate the source data into the common format to give to the data warehouse for storage.

The normalization of field data is related to field cleaning. By *normalization* of the data fields, we mean using a common format for all data belonging to the same type in order to make the comparison of fields possible. An example of string normalization is its conversion to capital letters; dates are said to be normalized if all of them follow the "dd/mm/yyyy" format. Other types of normalization can be thought in order to facilitate the comparison of equivalent fields, such as grouping words into phrases, correcting line breaks that separate a word, or using stemming techniques to keep common roots of words.

4.3.2 Special-Purpose Cleaning

When the domain of data to be cleaned is specific, for instance pharmaceutical names, or when the fields to clean belong to a specific application domain (for instance, name and address fields) special-purpose cleaning tools (also called *light-cleaning* tools in [Hurw97]) are able to solve common anomalies. Such cleaning tools use look-up tables to search for valid data (e.g., US mailing addresses) and dictionaries to look for synonyms and abbreviations (e.g., "St." and "Street") for the data in the fields. This way, corrections of spelling and validation of domain-specific information are obtained.

Due to their restricted domain of application, these tools perform very well. Some examples are: PostalSoft ACE, SSA (Search Software America), PostalSoft Library and Mailers +4. Some data warehouse-related products such as Carleton's Pure Integrate [PuIn98] (formerly known as Enterprise/Integrator) and ETI Data Cleanse [ETI98] also offer table-driven cleaning capabilities.

4.3.3 Domain-Independent Cleaning

In contrast with the last two sections, let us suppose from now on that the data to be cleaned is the result of a combination of data from heterogeneous data sources. The additional problem of having the same entity described by two different values may arise and has to be dealt with. To merge records that may be described by alternative values, approximate joins must be used in order to avoid losing connections between the records [Shas98]. In addition, the results obtained can be applied to determine, in terms of the database schemas, which attributes refer to the same category of entities.

The algorithms described by Monge and Elkan [MoEl96] are based on the principle of defining the degree of matching between two fields as the number of their matching atomic strings divided by their average number of atomic strings. Two strings are said to match if they are equal or if one is a prefix of the other. This paper describes three field-matching algorithms (basic, recursive, and Smith-Waterman) with different time complexities. The recursive algorithm, which is based on the partition of each field in subfields that are then matched with each other, is applied in an online searching tool called WebFind.

Carleton's Pure Integrate [PuIn98] product supports key-based (when records are identified by noncorrupted keys) and no-key-based matching (also called "fuzzy matching") to compare possible dirty records from different sources.

4.3.4 Rule-Based Cleaning

Another set of methods is used to detect field matching when merging data sources. The rule-based methods, besides using results from syntactical analysis, take into account a set of rules that establish equivalencies between records of different databases, taking into account the combination of several field matchings. These rules can be specified by the user or data warehouse builder or can be derived automatically by applying data mining techniques to the data sources.

4.3.4.1 User-Specified Rules

One example of user-specified rule system is the EDD Data Cleanser tool [EDD98] that uses a well-documented technology [HeSt95] to solve the "Merge/Purge" problem. The Merge/Purge problem arises whenever one wants to merge big volumes of data (that may be corrupted) from multiple sources as quickly as possible, and the resulting data is required to be as accurate as possible. Dirty data exists mainly because there were typographical errors or fraudulent activity leading to the existence of duplicate records about the same real entity. The method applied to eliminate duplicates and merge records is a sorted neighborhood method that involves first the creation of keys by analyzing each data source record, then sorting the records according to one of those keys, and finally merging matching records within a fixed-size window of records. The matching function used to compare data records is based on a set of rules (forming an Equational Theory) that establish correspondences between records. Two records match if they differ slightly by the application of a distance function. An excerpt example of these rules coded in C, as supplied in [HeSt98], is as follows:

Example 1: The goal is to merge records (*Person 1* and *Person 2*) with attributes: *Ssn, Name, Address, City, State, and Zip (code)*. Records compared belong to a fixed-size window.

```
for (all tuples under consideration) {
        for (the tuples inside the fixed-sized window) {
                Boolean similar-ssns = same-ssn-p(ssn1, ssn2)
                Boolean similar-names =
                        compare-names(name1, first-name1, last-name1,
                        initials-name1, name2, first-name2,
                        last-name2, initials-name2)
                if (similar-ssns and similar-names)
                        merge-tuples(person1, person2)
                Boolean similar-addrs = compare-addrs (street1, street2)
                Boolean similar-city = same-city(city1, city2)
                Boolean similar-zip = same-sip(zip1, zip2)
                Boolean similar-state = !strcmp(state1, state2)
                very-similar-addrs = (similar-addrs && similar-city &&
                        (similar-state || similar-zip));
                if ((similar-ssns || similar-names) && very-similar-addrs)
                        merge-tuples (person1, person2);
                ....
                }
        }
```

The sorted neighborhood method can appear in two more sophisticated forms. In the multipass approach, the basic algorithm is executed several times, each time using a different key for sorting, and the resulting records are obtained by the union of transitive closures over the intermediate results. Another approach, the duplicate elimination approach, eliminates exact or very close matching records during the sorting phase. This enhancement allows a reduction in time as the elimination of duplicates is done before the merging phase.

Two other examples that use a set of rules to guide the cleaning of data are Pure Integrate [PuIn98] from Carleton and the ETI Data Cleanse module [ETI98]. The former one allows the specification of merging rules based on several predefined criteria (for instance, choosing the most frequently occurring field value).

The main disadvantages associated to this kind of solution according to [Hurw97] are that writing rules is considered a time-consuming task, and those rules will never cover every possible anomaly in data. This last situation leads to exceptions that are handled manually by the user.

4.3.4.2 Automatically Derived Rules

Another set of tools that use rules to solve conflicts between records from different sources that describe the same real entity derive those rules automatically using data mining techniques. In fact, the contents of each data source are lexically analyzed and statistics involving words and relationships between them are found. Several data mining techniques, such as decision trees or associative rules, can be used to find data patterns. The result of such computation is a set of rules that govern each data source. An example of a commercial tool is WizRule [WizR98]. Some examples of database rules that result from applying data mining techniques are as follows:

Example 2:

mathematical rule:

A = B * C
WHERE
A = Total, B = Quantity, C = Unit Price
Rule's accuracy level: 0.99
rule exists in 1890 records

Accuracy level = ratio between the nb. of cases in which the formula holds and the total number of relevant cases.

if-then rule:

IF Customer IS "Summit"
AND Item IS Computer type X
THEN Salesperson = "Dan Wilson"

Rule's probability: 0.98
rule exists in 102 records
error probability < 0.1

Rule's probability = ratio between the nb. of records in which the conditions and the result hold and the nb. of records in which conditions hold with or without the result.

Error probability = chances that the rule does not hold in the entire population.

Another commercial tool that follows the same approach is Integrity [Vali98] from Vality. The major drawback of these approaches is the level of uncertainty that the derived rules imply.

4.3.5 Concluding Remarks on Data Cleaning

From the analysis of the cleaning techniques presented, we first conclude that many research issues have been addressed except in the case of multisource cleaning. In fact, Sects. 4.3.3 and 4.3.4 present some algorithms that determine the degree of matching between records from different sources, but the types of cleaning that involve exclusively one source are rather systematic and do not need innovative techniques.

A second aspect that is worth mentioning is the context of applicability of these cleaning techniques within the construction and maintenance of a data warehouse. In the introduction, we stated that the cleaning is done during data transformation. Yet, a distinction has to be made between the data transformations executed during the initial loading of the data from the operational data sources (to which we were actually referring) and the complementary data transformation that takes place during the data warehouse refreshment. In general, the data refreshment process updates the data warehouse according to changes in data sources. In terms of data cleaning, the obvious difference is the amount of data to be cleaned which is usually smaller in a refreshment situation. Fortunately, all the described data cleaning techniques can be used in loading as well as in refreshment. However, the strategy of applicability of those techniques is different in a refreshment process, in particular in what concerns the multisource data cleaning. In this case, the data warehouse is already composed of merged data and the arriving changes have somehow to be compared with the integrated and cleaned according to a given set of criteria. Moreover, changes from distinct data sources may not be detected all at the same time; consequently, no matching of all operational data sources can be done in order to discover the suitable data value for the corresponding integrated view. We can envisage several options for applying cleaning techniques that depend on the propagation time and data quality required. In our opinion, the study of the possible cleaning semantics and strategies to use during the refreshment of a data warehouse from several sources is an open area of research.

4.4 Update Propagation into Materialized Views

A data warehouse contains a collection of materialized views derived from tables that may not reside at the warehouse. These views must be maintained, i.e., they have to be updated in order to reflect the changes happening on the underlying tables. In this section, we shall focus on the problem of maintaining a collection of materialized views, i.e., the problem of computing the changes to be applied to the views and the actual application of these changes to the warehouse. An overview of research results related to the view maintenance problem is given in [GuMu95]. After introducing definitions and notations, we first present general results over the view maintenance problem and then we present results that are specific to the view maintenance in the data warehouse context.

4.4.1 Notations and Definitions

A *materialized view* V is a relation defined by a query Q over a set $\mathcal{R}$ of relations. We denote it as $V = Q(\mathcal{R})$. The *extent* of V, noted $\mathbf{V}$, is the bag of tuples returned by the query "select * from V." When the relations in $\mathcal{R}$ are updated, V becomes inconsistent. View refreshment is the process that reestablishes the consistency between R and V. This can be done within the transaction updating $\mathcal{R}$ (*immediate* refresh), or periodically, or when the view is queried (*delayed* refresh). The view can be refreshed using a full recomputation (*full refresh*) or a partial recomputation (view *maintenance*).

Let I and I' be two database states. Let $\Delta\mathcal{R}$ denote the difference between the instance of the relations of $\mathcal{R}$ in state I, denoted $I(\mathcal{R})$, and the instances in state I', denoted $I'(\mathcal{R})$. Let $\mathbf{V}$ be the extent of V in state I. Then V is *maintainable* if there exists an expression ΔQ such that, for any I and $\Delta\mathcal{R}$, $I'(V) = \Delta Q(\mathbf{V})$. V is *partially maintainable* if ΔQ is convenient only for certain modification operations of $\mathcal{R}$. For example, view V = min(select $A.1$ from A) is maintainable under insertions into A, it is not maintainable under deletions from A or updates of $A.1$. V is *self-maintainable* if ΔQ uses solely $\mathbf{V}$ and $\Delta\mathcal{R}$ for its refreshment. V is *weakly self-maintainable* if ΔQ uses also a set of relations $\mathcal{S} \subset \mathcal{R}$. (Weak) self-maintenance may be partial.

The maintenance of a view V may be decomposed in two steps: change computation and refresh. During the change computation step, the changes to $\mathbf{V}$, noted $\Delta\mathbf{V}$, are computed. During the refreshment step, $\mathbf{V}$ is updated by applying $\Delta\mathbf{V}$ to $\mathbf{V}$. More formally, we can write "ΔQ = compute ΔV; apply $\Delta\mathbf{V}$ to $\mathbf{V}$." In what follows, the expression that computes $\Delta\mathbf{V}$ is indicated as the differential expression of V. Let us remark that what we define "view maintenance" is often called "incremental view maintenance."

4.4.2 View Maintenance: General Results

In the sequel, we present research results that address the following issues:

- The characterization of views as (self) maintainable
- The derivation of differential expressions for these views
- The optimization of the maintenance process
- The maintenance of a collection of materialized views
- The performance of different refresh protocols

4.4.2.1 Characterizing (Self) Maintainable Views

There are two kinds of methods used in order to determine whether a view is (self) maintainable. *Static* methods provide criteria based on the pattern of queries and updates and possibly semantic knowledge such as functional dependencies. *Dynamic* methods test the (self) maintainability at run time.

Static criteria of maintainability or self maintainability for select-project-join (SPJ) views may be found in [BlLT86, CeWi91, ToBl88, GuJM96, QiWi91]. Criteria for SQL views with duplicates that use the union, negation, and aggregation operations may be found in [GuMS93, GrLi95]. Gupta et al. [GuMS93] also provide criteria for Datalog views with linear and general recursion. All these papers provide both static criteria and algorithms that take a (self) maintainable view and return its differential expression.

A run-time method is described in [Huyn96] where the author focuses on the problem of enhancing the self-maintainability of conjunctive views in the presence of functional dependencies. He provides an algorithm that generates the tests for detecting self-maintainability at run time and produces the differential formula to use if the self-maintainability test succeeds. In addition to providing (self) maintainability criteria, some work specified what additional data enable not (self) maintainable view to become (self) maintainable. For example, Gupta et al. [GuMS93] maintain counters in order to enforce the self maintainability of aggregate views. Another example may be found in [CGL*95] where the authors claim that, with a delayed refreshment protocol, it is necessary to store the changes to the operand relations in order to maintain the views.

In some sense the opposite problem to self-maintainability has been studied in [StJa96, StJa00]. Many OLAP client tools do not have database capabilities, i.e., they do not have the ability to perform relational queries and thus incremental view maintenance themselves. For such "externally materialized" views, a method is presented where the data warehouse computes the view differentials, translates them to the representation used by the client system, and sends them there.

4.4.2.2 Optimization

The problem of efficiently maintaining views has been studied along two dimensions. Local optimization focuses on differential computation, whereas transactional optimization improves the synchronization of the maintenance processing, the transactions querying the views, and the transactions updating the base relations.

Optimizing the differential computation of a view is the focus of [RoSS96]. The idea is to materialize subviews of a materialized view V in order to speed up the computation of ΔV. These additional views have to be maintained upon updates to the underlying base relations. The problem is to find the best balance between the cost incurred by the refreshment of the subviews and the benefit produced by using these views for computing ΔV. The set of subviews to materialize is computed by using a heuristics approach combined with a cost function.

The transactional optimization starts from the following facts: (1) the query transactions reading materialized views run concurrently with the update transactions modifying the base relations, and (2) with a *delayed* refreshment protocol, the maintenance is processed within the query transactions, while with an *immediate* protocol the refreshment is processed as part of the update transaction. The delayed protocol severely increases the query response time, and the immediate protocol makes update transactions long. The solution proposed in [CGL*96] consists in distributing the refreshment process among the update and query transactions.

For example, the update transaction may execute the propagation step and the query transaction the refreshment step. Another possibility may consist of processing the change computation in a separate transaction [AdGW97]: one maintenance transaction may handle the changes produced by several update transactions. The authors provide measurements showing that this solution greatly improves the refreshment cost with respect to the CPU loading cost.

In [QuWi97], the authors define an algorithm that avoids contention problems between maintenance transactions and query transactions. The trick consists in implementing a specific multiversion protocol. Every view has two versions: one is updated by the view maintenance process and the other is read by the users. The switching from a version to the other is performed at the end of the refreshment process. This is particularly useful for data warehouses which are queried 24 hours a day.

4.4.2.3 Joint Maintenance of a Set of Views

The refreshment of a set of views needs specific algorithms intended to efficiently schedule the propagation and refresh processing of each individual view with respect to the global refreshment process.

Hull and Zhou [HuZh96] present a rule-based solution for maintaining a set of views. The interdependency between base relations and views is represented by a graph called VDP. Active rules maintain materialized views by a breadth-first traversal of the VDP.

Maintaining a set of views raises the problem of mutual consistency of the views. [CKL*97] focus on the consistency problem that arises when several interdependent views are maintained with different refreshment policies. In order to achieve consistency, the views are grouped in a way that the views belonging to the same group are mutually consistent. The paper provides algorithms to compute the view groups and maintain them. The method works in a centralized environment. It is easily applicable to the data warehouse context by assuming that (1) the base relations are at the source level, and (2) the base relations belonging to the same group are mutually consistent, i.e., each update transaction running at the source level is reported in a single message.

Algorithms of [ZhWG97] the maintenance of a set of views, optimize in a consistent way, using a concurrent propagation of changes. Indeed, when a single data modification has to be mirrored in various views, the application of different ΔVs has to be performed in a single transaction. The algorithms use the knowledge of all the views that are related with a single base relation and refresh the views only when all the related ΔVs have been computed. Maintaining a set of views brings additional opportunities regarding optimization (e.g., by factoring common computations). Such a solution is proposed by [MuQM97] to optimize the maintenance of materialized aggregate views (also called *summary tables*) that are intensively used in OLAP applications. The authors claim that classical maintenance algorithms fail to efficiently maintain such views. The proposed solution consists of materializing additional "delta summary tables" intended to store the result of the propagation processing. A delta summary table lightly differs from a classical delta table associated with a given view. While the former stores the changes to the view, the latter stores aggregated data used to refresh several summary tables. This approach provides a gain in space and propagation time, an optimized availability of the views (the delta summary tables are maintained without locking the views) but is paid by a more complex refreshment computation.

4.4.2.4 Evaluating View Maintenance Algorithms

View maintenance algorithms have been evaluated in two ways. One compares the performances of full refresh with respect to (incremental) maintenance. The other compares the performance of maintenance algorithms applied at different delays with respect to the modifications of $\mathcal{R}$.

It is generally accepted that maintaining a view is more efficient than recomputing the view as long as the amount of data modified in the underlying base relations is sufficiently small compared to the size of these relations. In [CKL*97], the authors study the threshold where a complete recomputation becomes more efficient for SPJ views. The experiments point out the importance of the transaction patterns. For example, with transactions executing only one kind of modification (e.g., only insertions), the incremental maintenance under insertions defeats full recomputation as long as base tables have become about 23% larger. The measures for update and delete give 7% and 15% respectively. Similar results are given in [Hans87].

Hanson [Hans87], Adelberg et al. [AdKG96], and Colby et al. [CKL*97] compare refreshment protocols by measuring the query response time, the overhead for transactions modifying $\mathcal{R}$, the percentage of transactions reading inconsistent data, the number of refresh operations. The measurements are expressed in terms of various parameters (for example, [AdKG96] varies the number of relations involved in the definition of the view).

4.4.3 View Maintenance in Data Warehouses – Specific Results

The maintenance of the views stored in a data warehouse raises specific consistency and optimization problems. In what follows, we present the main research results handling these problems.

4.4.3.1 Consistency

Informally the consistency of a view V with respect to a certain maintenance protocol, describes the relationship between a temporal sequence of source states and a (temporal sequence of) V extents resulting from the maintenance process.

The consistency of views has been first studied in the context of maintenance algorithms based on queries against the sources (called *nonautonomous view maintenance*). [ZhGW96] defines the consistency with respect to one source, and [ZGHW95] and Agrawal et al. [AESY97] extend this definition for several remote sources. Proposed levels of consistency are as follows:

- *Convergence* where $V = Q(\mathcal{R})$ only once that all maintenance activities at the data warehouse have ceased
- *Weak consistency* where, for every extent V, there is a given state of source(s) $\mathcal{R}$ such that V is related to $V = Q(\mathcal{R})$
- *Strong consistency* that ensures weak consistency and where the temporal sequence of V preserves the order of temporal sequence of source(s) states
- *Completeness* where there is a complete order-preserving mapping between the states of the view and the states of the source(s)

As pointed out in [ZhGW96], different consistency levels may be required by different data warehouse applications. Zhou et al. [ZhGW96], [ZGHW95], and Agrawal et al. [AESY97] define nonautonomous view maintenance algorithms achieving strong or even, in [AESY97], complete consistency. The algorithms work as follows. When a change is signaled at the data warehouse, the sources are queried in order to compute the differential expression for this change. Due to concurrent updates, the answer may contain errors. The algorithms implement a process compensating these errors.

Until recently, commercial view maintenance techniques provided for weak consistency only. However, advances in the integration of indexing techniques for star schemas and aggregates with view maintenance techniques in products such as Red Brick Vista [BCC*01] show that practice is rapidly adopting these important performance improvements.

The consistency of views has received other definitions as well. Baralis et al. [BaCP96] define the consistency of views in the context of autonomous view maintenance and distributed transactions over sources related by integrity constraints. The authors use a notion of weak consistency corresponding to a consistency level where the data of the views are correct with respect to the defined integrity constraints but may reflect a state of source(s) that has never existed. The authors develop algorithms ensuring strong or weak consistency. For example, if the base relations, noted $\mathcal{B} \subset \mathcal{R}$, of some source $S1$ are not involved in any integrity constraints over sources, then the maintenance of V with respect to atomically performed changes to $\mathcal{B}$ provide a *weakly* consistent V. This result leads to faster spreading of changes of $\mathcal{B}$ to V. Indeed, V may be maintained as soon as the changes have been received at the data warehouse. Hull and Zhou [HuZh96] give definitions of consistency applying to algorithms that mix autonomous and nonautonomous view maintenance.

4.4.3.2 Optimization

In order to optimize the view maintenance process, several research works emphasize on making the data warehouse globally self-maintainable. The idea is to store sufficient data inside the warehouse to allow the maintenance of the views without accessing remote relations. The problem, however, is to minimize the amount of data needed to self-maintain the views. Quass et al. [QGMW96] treat the case of SPJ views. They take benefit of key attributes and referential integrity constraints to minimize the amount of additional data. Additionally, the authors define the related modifications of differential expressions. Huyn [Huyn97] provides algorithms that test view maintainability in response to base updates, based on the current state of all the views in the warehouse. Garcia-Molina et al. [GaLY98] propose to reduce the amount of data by periodically removing materialized tuples that are no longer needed for computing the changes to the high-level views.

4.4.3.3 Temporal Data Warehouses

In [YaWi97], the authors address the issue of *temporal data warehouse* where the views reflect a chronicle of source data, and sources are able to manipulate only the current information. They developed a temporal data model and a corresponding algebra. In order to avoid "reinventing the wheel," the authors reduce the temporal relations and operators to their nontemporal counterparts. Thus, the techniques developed in nontemporal proposals remain convenient. Two crucial facts have been pointed out. First, the views have to be self-maintainable since the past data are not available at source level. Second, the view refreshment may be induced also because of simple time advancing. For performance reasons, this kind of view refreshment is performed only at query time.

4.5 Toward a Quality-Oriented Refreshment Process

The refreshment process aims to propagate changes raised in the data sources to the data warehouse stores. This propagation is done through a set of independent activities (extraction, cleaning, integration, ...) that can be organized in different ways, depending on the semantics one wants to assign to the refreshment process and on the quality one wants to achieve. The ordering of these activities and the context in which they are executed define the semantics and influence this quality. Ordering and context result from the analysis of view definitions, data source constraints and user requirements in terms of quality factors.

The refreshment process is an event-driven system which evolves frequently, following the evolution of data sources and user requirements. There is no refreshment strategy that is suitable for all data warehouse applications or the whole data warehouse lifetime. Besides the methodology and tools which support the definition and implementation, our contribution is also to provide quality factors that validate whether a given refreshment process meets the user requirements.

In this section we describe different semantic features and quality factors that affect the refreshment process.

4.5.1 Quality Analysis for Refreshment

The semantics of the refreshment can be defined as the set of all design decisions that contribute to provide relevant data to the users, while at the same time, fulfilling all the quality requirements.

We have already described a data warehouse as layers of materialized views on top of each other. Yet a view definition is not sufficient to capture the semantics of the refreshment process. Indeed, the query which defines a view does not specify whether this view operates on a history or not, how this history is sampled, whether the changes of a given source should be integrated each hour or each week, and which data timestamp should be taken when integrating changes of different sources. The view definition does not include specific filters defined in the cleaning process, such as choosing the same measure for certain attributes, rounding the values of some attributes, or eliminating some confidential data. Consequently, based on the same view definitions, a refreshment process may produce different results depending on all these extra-parameters which have to be fixed independently, outside the queries that define the views.

4.5.1.1 Quality Dimensions

Quality dimensions can be considered as property types which can characterize any component of the data warehouse (sources, ODS, views). We define hereafter four quality dimensions that best characterize the refreshment process.

- *Data coherence.* An example is choosing the right timestamp for the data involved in a join between different sources. Depending on the billing period and the extraction frequency, the extracted data may represent only the last three or four months. As another example, the conversion of values to the same measurement unit allows also to do a coherent computation.
- *Data completeness.* An example is checking whether the data acquired from the sources answer correctly the query which defines the view. If there is a representative sample of data for each view dimension we can check whether the extracted values from the billing sources of each country provides 10% or 100% of the whole number of rows. The completeness of the sources determines also the *accuracy* of the computed values in the views.
- *Data accuracy.* Defines the granularity of data provided by the sources or computed in the views. Some billing sources may have only a few clients who have adopted a detailed billing, others have only one item for each period and for each client. The data in the source may be considered as complete but not necessarily usable with respect to the view definition.
- *Data freshness.* In the context of intensive updates and extraction, there might be a problem to find a tradeoff between accuracy and response time. If a view wants to mirror immediate changes of sources, the data warehouse administrator should first negotiate the right availability window for each source, and then find an efficient update propagation strategy whose cost is less than the time interval between two successive extractions.

4.5.1.2 Quality Factors

For each quality dimension, several quality factors may be defined in a data warehouse. For example, one can define the accuracy of a source, the accuracy of a view, the completeness of a source content, the completeness of the ODS, or the completeness of the source description, etc. However the quality factors are not necessarily independent of each other, e.g., completeness and coherence may induce a certain accuracy of data. We distinguish *primary* and *derived* quality factors. For example, the completeness of a source content may be defined with respect to the reality this source is supposed to represent. Hence, its completeness is a subjective value directly assigned by the data warehouse administrator. On the other hand, the completeness of the ODS content can be a formula over the completeness of the sources. The definition of all these quality factors constitutes the user requirements that will be taken into account to define a specific policy for the refreshment process. It is obvious that this policy evolves with the user's needs without necessarily changing queries which define the data warehouse views.

Quality factors allow the evaluation of the design decisions and check whether they fit user requirements or not. In the sequel, we mention several quality factors which are relevant to the refreshment process. Some of them are arbitrarily assigned by the data warehouse administrator; others are computed from the former.

- Availability window of a source (defined as an authorized access frequency and a duration of availability);
- Frequency of data extraction from a source, of multiple source data integration, and of update propagation;
- Estimated response time of each algorithm which implements a refreshment activity (extraction, integration, update propagation, etc.); we can assume that this response time include computation time and data transfer time;
- Expected response time for each refreshment activity;
- Estimated volume of data extracted each time from each source;
- Total duration of the history (for which period of time the history is defined);
- Actual values of data freshness;
- Expected value of data freshness.

4.5.1.3 Design Choices

The evaluation of the values of these parameters will be based on design policies which can evolve with the semantics of the refreshment process. Underlying the policies are techniques, i.e., rules, events and algorithms which implement the strategies on which refreshment activities are based.

- The *granularity of the data* if different levels of details are given in the same source or different sources, or in any other data store of the data warehouse;
- The *time interval to consider in the history* of each source or the history of the ODS, i.e., the chunk of data which is of interest to a given query;
- The *policy of data extraction and cleaning,* choice of extraction frequency, interaction between extraction and cleaning, or choice of triggering events;
- The *policy of data integration,* i.e., how to join data with different timestamps, when to integrate extracted data, or how to consider redundant sources;
- The *policy of update propagation,* i.e., incremental update or complete recomputation, or when to trigger update propagation.

4.5.1.4 Links between Quality Factors and Design Choices

Quality dimensions, quality factors, and design choices are closely related. The table in Fig. 4.4 shows some possible links among them.

Quality dimension	DW objects	Derived quality factors	Primary quality factors	Design choices
Coherence	• Sources • ODS • Views	• Extraction frequency of each source • Estimated response time of extraction for each source, of integration and of update propagation	• Availability window of each source • Expected response time for a given query	• Granularity of data • Extraction and cleaning policy • Integration policy
Completeness	• Sources • ODS	• Extraction frequency of each source	• Availabilit window of each source • History duration for each DW store	• Extraction policy • Integration policy
Accuracy	• Sources • ODS • Views	• Extraction frequency of each source	• Availability window of each source • History duration for each DW store	• Granularity of data • Time interval in the history • Extraction policy • Integration policy
Freshness	• Sources • ODS • Views	• Extraction frequency of each source • Actual freshness for a given query • Actual response time for a given query	• Availability window of each source • Expected freshness for a given query • Estimated response time of extraction for each source, of integration and of propagation • Volume of data extracted and integrated	• Extraction policy • Integration policy • Update policy

Fig. 4.4. Link between quality factors and design choices

As we have seen, the refreshment process depends on various quality factors given as explicit requirements of user applications. We have also listed some of the design decisions which help achieve these requirements. Before starting the design process, it would be interesting to check whether quality factors are coherent with each other, and possibly refine them by a negotiation process among the data warehouse administrator, the source administrators, and the users. One of the important issues during the design process is to validate whether actual design choices satisfy the quality requirements. This validation may lead to one of the following results:

- Change of the refreshment strategy or techniques.
- Negotiation with sources administrators to get more accessibility and availability on the sources or with the users to downgrade some of their quality requirements.

The steps to deal with these validation problems are as follows:

- Find a computation procedure which derives, from source constraints (or quality factors) and from the estimated performance of the refreshment techniques, actual quality values that can be achieved,
- Make a confrontation between quality values expected by applications and quality values actually provided by data sources and refreshments techniques and identify the main mismatches.

4.5.2 Implementing the Refreshment Process

This section describes the way to implement the refreshment process. First, it shows how the refreshment process can be modeled as a workflow. Second, it shows how this workflow can be implemented by active rules.

4.5.2.1 Planning the Refreshment Process

The refreshment process should be defined by planning its workflow with respect to the design choices derived from the desired quality factors. As for the integration process [CDL*97], it is possible to view the refreshment process through different perspectives:

- *Client-driven refreshment* describes part of the process that is triggered on demand by the users. This part mainly concerns update propagation from the ODS to the aggregated views. The on-demand strategy can be defined for all aggregated views or only for those for which the freshness of data is related to the date of querying.
- *Source-driven refreshment* defines part of the process that is triggered by changes made in the sources. This part concerns the preparation phase (Fig. 4.1). The independence between sources can be used as a way to define different preparation strategies, depending on the sources. Some sources may be associated with cleaning procedures, others not. Some sources need a history of the extracted data, others not. For some sources, the cleaning is done on the fly during the extraction, for some others after the extraction or on the history of these changes. The triggering of the extraction may also differ from one source to another. Different events can be defined, such as temporal events (periodic or fixed absolute time), after each change detected on the source, on demand from the integration process.
- *ODS-driven refreshment* defines part of the process that is automatically monitored by the data warehouse system. This part concerns the integration phase and may be triggered at a synchronization point, defined with respect to the ending of the preparation phase. Integration can be considered as a whole and concerns all the source changes at the same time. In this case, the refreshment can be triggered by an external event which might be a temporal event or the ending of the preparation phase of the last source. The integration can also be sequenced with respect to the termination of the preparation phase of each source, that is, the extraction is integrated as soon as its cleaning is finished. The ODS can also monitor the preparation phase and the aggregation phase by the generation of the relevant events that triggers activities of these phases.

In the very simple case, both approaches are synchronized and form one single strategy. In a more complex case, there may be as many strategies as the number of sources or high-level aggregated views. In between, there may be, for example, four different strategies corresponding to the previous four phases. The strategy to choose depends on the semantic parameters and also on the tools available to perform the refreshment activities (extraction, cleaning, integration). Some extraction tools perform the cleaning on the fly while some integrators propagate immediately changes until the high level views. Then, the workflow in Fig. 4.2 is a logical view of the refreshment process. It shows the main identified activities and the potential event types which can trigger them.

We can distinguish several event types which may trigger the refreshment activities. Figure 4.5 summarizes some of these events for each activity as well as examples when necessary.

Activity	Event types	Examples
Customization	• After update propagation termination • At the occurrence of a temporal event • At the occurrence of an external message sent by the DWA or an application program	• Before each query evalua tion
Update propagation	• After termination of the integration phase • At the occurrence of a temporal event • At the occurrence of an external message sent by the DWA or an application program	• End of integration or end of ODS archiving • Before customization
History management for ODS	• After termination of integration	
Data integration	• After termination of the preparation phase of each source • After termination of the preparation phase of all sources • At the occurrence of a temporal event • At the occurrence of an external message sent by the DWA or an application program	• End of extraction or end of cleaning or end of archiving • At a predefined synchronization point • Every day at 5, every week • Before update propagation
History management of source data	• Termination of extraction • Termination of cleaning	
Data cleaning	• Termination of extraction • Termination of history management	
Data extraction	• After each change on the source data • After termination of each transaction executed on a data source • At the occurrence of a temporal event • At the termination of a refreshment activity • At the occurrence of an external message sent by the DWA or an application program.	• Insert, delete, or update • Every two hours, every Monday at 5, every ten committed transactions • End of archiving, end of cleaning, end of integration • Before integration

Fig. 4.5. Event types that trigger refreshment activities

Termination covers different event types, depending on the activity ending (by commit or by abort for example). Synchronization points can be defined in different ways: if there is any ordering on the sources, the synchronization point can be defined as the commit of the preparation phase of the last source. It can also be defined as the commit of the preparation phase of the first two sources. A synchronization point can also be defined before each binary operator involving data of different sources.

Figure 4.6 gives a possible workflow for the motivating example defined earlier. We can define another refreshment scenario with the same sources and similar views. This scenario mirrors the average duration and cost for each day instead of for the past six months. This leads to change the frequency of extraction, cleaning, integration, and propagation.

Within the workflow, which represents the refreshment process, activities may be of different origins and different semantics. The refreshment strategy is logically considered as independent of what the activities actually do. However, at the operational level, some activities can be merged (e.g., extraction and cleaning), and some others decomposed (e.g., integration).

Activities of the refreshment workflow are not executed as soon as they are triggered, since they may depend on the current state of the input data stores. For example, if the extraction is triggered periodically, it is actually executed only when there are effective changes in the source log file. If the cleaning process is triggered immediately after the extraction process, it is actually executed only if the extraction process has gathered some source changes. Consequently, we can consider that the state of the input data store of each activity may be considered as a condition to effectively execute this activity.

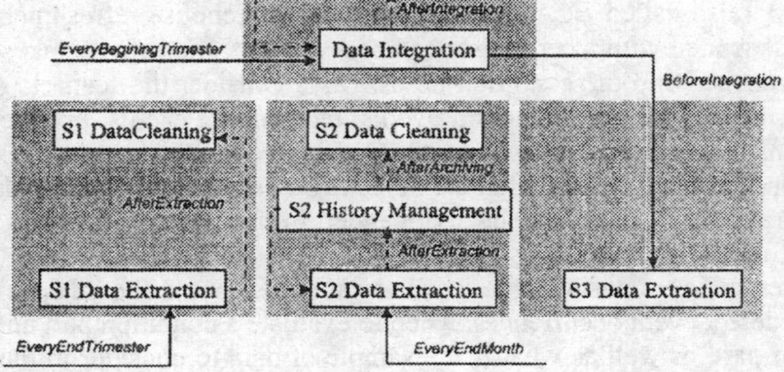

Fig. 4.6. First example of refreshment scenario

There may be another way to represent the workflow and its triggering strategies. Indeed, instead of considering external events such as temporal events or termination events of the different activities, we can consider data changes as events. Hence, each input data store of the refreshment workflow is considered as an event queue that triggers the corresponding activity. Figure 4.2 is sufficient to describe this approach, we have just to consider that the data stores as queues of events which are produced by corresponding activities. However, to be able to represent different refreshment strategies, this approach needs a parametric synchronization mechanism which triggers the activities at the right moment. This can be done by introducing composite events which combine, for example, data change events and temporal events. Another alternative is to put locks on data stores and remove them after an activity or a set of activities decide to commit. In the case of a long term synchronization policy, as it may happen in data warehouses, this latter approach is not sufficient.

4.5.3 Workflow Modeling with Rules

The previous example has shown how the refreshment process can depend on some parameters, independently of the choice of materialized views. Moreover, as stated before, the refreshment strategy is not defined once for all; it may evolve along with the user needs, which may result in the change of the definition of materialized views or the change of desired quality factors. It may also evolve when the actual values of the quality factors lower with the evolution of the data warehouse or the technology used to implement it. Consequently, in order to master the complexity and the tendency of the data warehouse to evolve, it is important to provide a flexible technology that allows the accommodation of these features.

It is argued in many papers that active rules provide interesting features which, at some extent, make possible this flexibility. It is obvious to show that the refreshment workflow, whatever is the way of describing its strategy, can be easily implemented as a set of active rules executed under a certain operational semantics.

Active rule systems provide a syntax and semantics which describe an application as a set of rules of the following form: *On <event-type> If <condition> Then <action>* (also called *ECA* rules). In the data warehouse refreshment context, a rule is associated with a certain refreshment activity: the activity takes place if the event happened and the condition holds. If we consider the scenario examples of Fig. 4.6, event types are temporal events, termination events, or external events, conditions are test expressions over the updates in the data repositories (for example, a condition may specify that the *S2 DataCleaning* activity takes place solely if a certain amount of data has been appended in the History), actions are refreshment activities to execute.

The semantics under which a scenario can be executed should specify how and when to detect event occurrences, when to evaluate a condition part and to execute an action part, as well as what is the sample of data to consider in the history for the condition and the action parts. Detecting occurrences of events triggering the rules generally requires to detect primitive events and to combine them. Take for example, the event types described in Fig. 4.5. Then detecting the event "the preparation phase of all sources is terminated" requires to detect the termination of the preparation phase of each source taken individually and then to combine these primitive events. So, specifying how and when to detect occurrences of the events triggering the rules requires first to specify how and when to detect the underlying primitive events, and second to specify how and when these primitive events have to be combined. In the context of workflows, activities are considered as atomic, that is, black boxes which have no meaning for the workflow, except that they terminate by generating certain event occurrences. The semantics of refreshment activities, e.g., the queries which extract data, the cleaning rules, and the integration rules, are not meaningful for the global refreshment strategy, unless one wants to write these activities as active rule applications. In this case, the semantics of the scenario should be considered as a different problem which is not handled by the same instance of the active system.

However, as discussed in [BFL*97], most of the research prototypes and relational products offering active functionalities today are "prefabricated systems," which means that they offer a predefined active language and execution model, although with a wide range of capabilities. Recent studies have showed that an application developer often encounters two main difficulties when wanting to use a prefabricated active system to develop real-life applications: (1) the language and the execution semantics offered by the system may not match the needs of the applications, and (2) the rule execution engine offered by the active system may not provide performance which is good enough for the needs of the application. The ideal active system for the data warehouse administrator or developer is an open system which allows one to adapt an implementation to the application needs, as often as the latter ones evolve. This is the approach we have followed [BFM*98] and adopted for the refreshment process. We designed and implemented modular system components that provide functionalities that are essential to engineer a customized and efficient active rule system. Thus, instead of providing the user with a prefabricated system, we advocate a toolkit-based approach that is aimed to facilitate the development of a rule execution engine and event detectors, according to the needs of an application.

4.5.3.1 Main Features of the Toolkit

- The toolkit is independent of any ECA rule language, i.e., it does not impose any particular ECA rule language. In fact, the toolkit does not provide any specific code to compute the detection of complex events, evaluate rule conditions, or execute rule actions. However, it enables implementing a very large variety of rule execution models.
- The toolkit is independent of any database system, i.e., the toolkit does not assume the use of any predetermined database system. The requirements put on the database system are minimal. In particular, it can be a passive system. In its current implementation, the code resulting from the use of the toolkit is a rule execution engine that interfaces between a database system and an application program using a Java Data Base Connection (JDBC) protocol.
- The organization of the toolkit is based on a logical model of an active system that consists of the specification of four main functional units: event detector, event manager, reaction scheduler, and execution controller, as well as a communication model among these four units. Different communication models are possible depending, for instance, on the number of units that can be executed concurrently, the type of communication between two units (whether synchro-

nous or asynchronous), etc. Each communication model entails restrictions on the possible active rule execution semantics and forces a specific behavior of the functional units. Hence, the logical view of the toolkit suggests a methodology to specify the type of rule execution model to be implemented.

- The toolkit enables a scalable and adaptable implementation of each of the functional units that respects the specifications made at the communication model level. To this aim, the toolkit provides a hierarchy of classes for each functional unit that enables the developer to select and specialize the classes that are needed to implement the desired execution model. As a result, the rule execution engine will only integrate the functionalities necessary to fulfill the gap between the needs of the application and the capabilities of the database system.

4.5.3.2 Functional Architecture of the Active Refreshment System

The toolkit is devoted to specify and implement a data warehouse refreshment system having the architecture shown in Fig. 4.7. The system is composed of four functional units (gray boxes): event detector, event monitor, rule scheduler, and execution controller. These units exchange data (thin arrows) and are controlled by synchronization mechanisms (bold arrows). We first describe the role of each functional unit taken individually, with respect to the data flow in the system: what data are produced, what data are handled, and what the treatments are. Then we elaborate over the control flow.

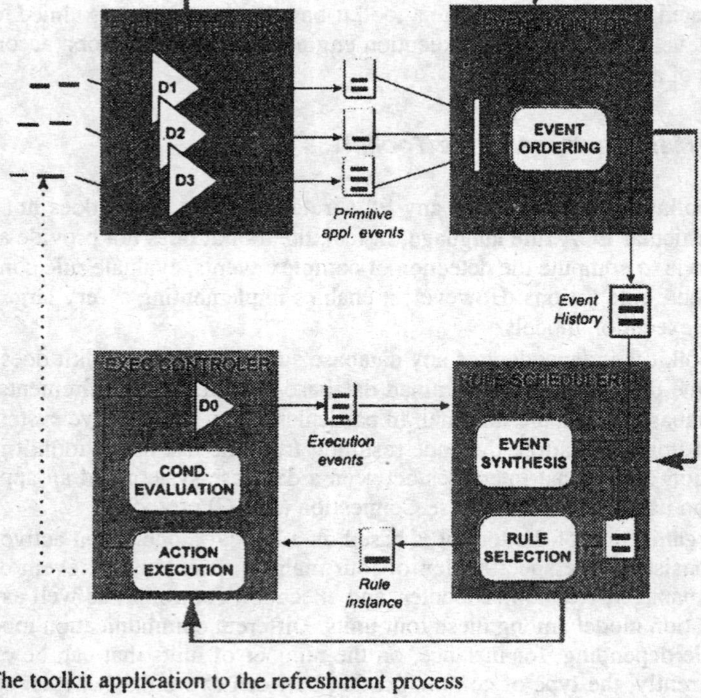

Fig. 4.7. The toolkit application to the refreshment process

The event detection function is in charge of implementing the mechanisms needed to detect the primitive events (i.e., to trap and to announce them when they occur). The detection mechanisms are distributed among a set of event detectors. The event monitor function takes the primitive events produced concurrently by the detectors and stores them in an event history with respect to an event ordering policy (for example, it may apply static or dynamic priorities between the detectors, or handle all the events as a whole). The execution controller function monitors the execution of the rule instances. Given a rule instance r_i, the evaluation of its condition and the execution of its action are respectively controlled by the condition evaluation component and the action execution component. By evaluation and execution control, we mean computation launching and result reporting (i.e., condition true or false, action terminated, aborted, etc). The results are transferred to the rule scheduler in the form of execution events. Depending on the needs of the application and on the underlying system capabilities, it is possible to envision a concurrent execution of several rule instances. The rule scheduler function schedules the reaction to event occurrences. The event synthesis component computes the triggered rules and produces the corresponding rule instances. The rule selection component operates over these rule instances and selects the rule instances to evaluate or to execute, given the reporting produced by the execution controller. The semantics of both the event synthesis and rule selection mechanisms is application dependent.

The control flow of the system depends both on the application needs and on the refreshment strategy. Specifying the control flow for a given refreshment system requires a decision on how many processes may run concurrently and the points where concurrent processes have to be synchronized. More precisely, the specification of the control flow consists in describing the concurrency policy between detectors and event monitor, event monitor and rule scheduler, and rule scheduler and execution controller. The choice of synchronization options impacts the execution semantics. For example, suppose that the event monitor cannot run concurrently with the rule scheduler. Then, during the operation of the rule scheduler, the event history state cannot change, so no rules are newly triggered.

4.6 Implementation of the Approach

In this chapter, we have discussed details of the continuous data refreshment process needed to keep a data warehouse useful. We have noted an enormous variety of needs and techniques where it is easy to get lost in the details. Therefore, in the last section, we have shown how to support this process with a comprehensive toolkit based on active database technology.

The DWQ framework from Sect. 2.7 provides different perspectives and abstraction levels that clearly define all data warehouse objects. The DWQ framework is based on a generic repository that provides meta classes that should be specialized, instantiated, and overridden in order to define the object types which characterize a specific data warehouse application.

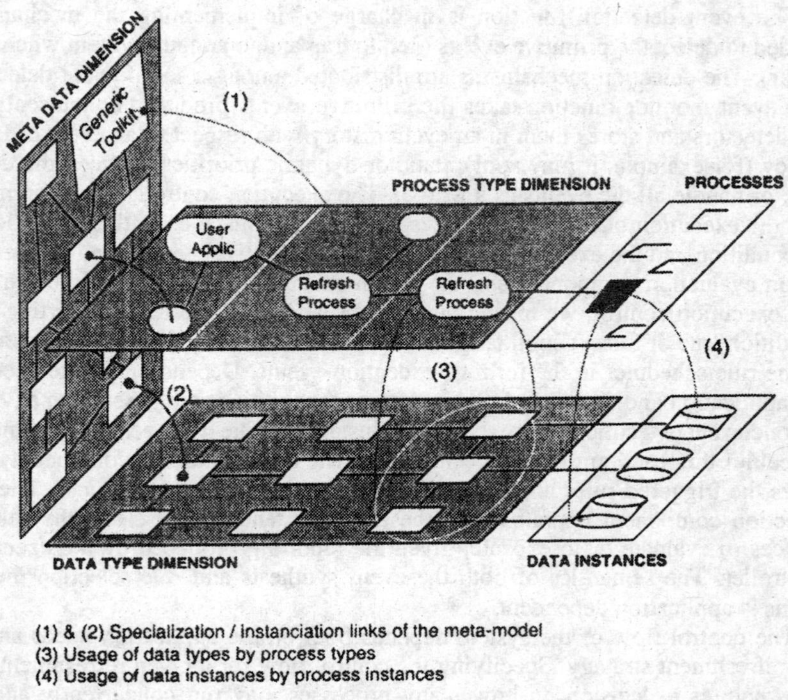

Fig. 4.8. Refreshment process within DWQ framework

Within this framework, the refreshment process is a component of the Process Dimension (Fig. 4.8). At the conceptual level, it is seen as an aggregation of different activities that are organized into a generic workflow. Component activities use metadata of the Data Type Dimension, such as source schema descriptions, ODS, or views. The workflow also uses its specific metadata such as event types, data stores, activity descriptions, and quality factors. Each component is specified at the conceptual, logical, and physical levels.

Based on these ideas, several research prototypes for refreshment have been built with a strong focus on data cleaning. INRIA's *AJAX* [GFSS00] deals with typical data quality problems, such as *the object identity problem, errors due to mistyping,* and *data inconsistencies* between matching records. This tool can be used either for a single source or for integrating multiple data sources. AJAX provides a framework wherein the logic of a data cleaning program is modeled as a directed graph of data transformations that start from some input source data. Four types of data transformations are supported:

- *Mapping transformations* standardize data formats (e.g., date format) or simply merge or split columns in order to produce more suitable formats.
- *Matching transformations* find pairs of records that most probably refer to same object. These pairs are called *matching pairs,* and each such pair is assigned a similarity value.

- *Clustering transformations* group together matching pairs with a high similarity value by given grouping criteria (e.g., by transitive closure). *Merging transformations* are applied to each individual cluster in order to eliminate duplicates or produce new integrated data source.

AJAX also provides a declarative language for specifying data cleaning programs, which consists of SQL statements enriched with a set of specific primitives to express mapping, matching, clustering, and merging transformations. Finally, an interactive environment is supplied to the user in order to resolve errors and inconsistencies that cannot be automatically handled and support a stepwise refinement design of data cleaning programs. The theoretic foundations of this tool can be found in [GFSS99], where apart from the presentation of a general framework for the data cleaning process, specific optimization techniques tailored for data cleaning applications are discussed.

Similarly, the *Potter's Wheel* system [Rahe00] offers algebraic operations over an underlying data set, including *format* (application of a function), *drop, copy, add* a column, *merge* delimited columns, *split* a column on the basis of a regular expression or a position in a string, *divide* a column on the basis of a predicate (resulting in two columns, the first involving the rows satisfying the condition of the predicate and the second involving the rest), *selection* of rows on the basis of a condition, *folding* columns (where a set of attributes of a record is split into several rows), and *unfolding*. Optimization algorithms are provided for CPU usage of certain classes of operators.

ARKTOS [VVS*01] supports practical data cleaning scenarios by providing explicit primitives for the capturing of common tasks via a uniform metamodel for ETL processes. ARKTOS provides three ways to describe an ETL scenario: a graphical point-and-click front end and two declarative languages which are variants of XML and SQL, respectively.

5 Multidimensional Data Models and Aggregation

This chapter is devoted to the modeling of multidimensional information in the context of data warehousing and knowledge representation, with a particular emphasis on the operation of *aggregation*.

The current information technology expectations are to help the knowledge worker (executive, manager, analyst) make more effective and better decisions. Typical queries that a knowledge worker would like to make to its enterprise knowledge repository – the data warehouse – are the following:

- What were the sales volumes by region and product category for the last year?
- How did the share price of computer manufacturers correlate with quarterly profits over the past 10 years?
- Which orders should we fill to maximize revenues?
- Which of two new medications will result in the best outcome: higher recovery rate and shorter hospital stay?
- How many disk drives did we ship to the eastern region last quarter in which the quantity shipped was greater than 10, and how much profit did we realize from those sales as opposed to those with less than 10?

It is clear that such requirements cannot easily be fulfilled by traditional query languages. In the following, we will survey the basic concepts of data modeling which are the foundations of commercial data warehousing systems, and the operations allowed on the data. Main sources for this survey are [AbGr95, ChDa96, Coll96, Deje95, Fink95, Fink96, OlCo95, Rade95, Rade96].

Before proceeding, we roughly compare the way in which classical databases are used with the way in which data warehouses are used. The traditional market of databases deals with online transaction processing (OLTP) applications. OLTP applications consist of a large number of relatively simple transactions. The transactions usually retrieve and update a small number of records that are contained in several distinct tables. The relationships between the tables are generally simple. For example, a typical OLTP transaction for a customer order entry might retrieve all of the data relating to a specific customer and then insert a new order for the customer. Information is selected from the customer, customer order, and detail line tables. Each row in each table contains a customer identification number which is used to relate the rows from the different tables. The relationships between the records are simple, and only a few records are actually retrieved or updated by a single transaction.

In contrast to OLTP applications of databases, data warehouses are designed for online analytical processing (OLAP) applications. OLAP applications are quite different from OLTP applications. OLAP is part of decision support systems (DSS) and of executive information systems (EIS). OLAP functionality is characterized by dynamic multidimensional analysis of consolidated enterprise data that supports end user analytical and navigational activities. At the practical level, OLAP always involves interactive querying of data, following a thread of analysis through multiple passes, such as "drill-down" into successively lower levels of detail. The information is "multidimensional," meaning for the user that it can be visualized in grids. Information is typically displayed in cross-tab reports, and tools provide the ability to pivot the axes of the cross-tabulation. We will consider in this chapter the parts of these operations that are always read-only ("narrow OLAP"). A characteristic of OLAP tools for data analysis is to allow the consolidation of data to higher levels while still supporting queries down to the detailed level.

As an example of an OLAP analysis session, consider the following. An OLAP database may consist of sales data which has been aggregated by *Region*, *Product type*, and *Sales channel*. A typical OLAP query might access a multigigabyte/multiyear sales database in order to find all product sales in each region for each product type. After reviewing the results, an analyst might further refine the query to find the sales volume for each sales channel within region/product classifications. As a last step, the analyst might want to perform year-to-year or quarter-to-quarter comparisons for each sales channel.

In the following, we give an intuition on how these transactions can be realized. Relational database tables contain records (or rows). Each record consists of fields (or columns). In a normal relational database, a number of fields in each record (keys) may uniquely identify each record. In contrast, the multidimensional data model is an n-dimensional array (sometimes called a "hypercube" or "cube"). Each dimension has an associated hierarchy of levels of consolidated data. For instance, a spatial dimension might have a hierarchy with levels such as country, region, city, office. In the example of Fig. 5.1, chosen dimensions are *Product, Region,* and *Month*.

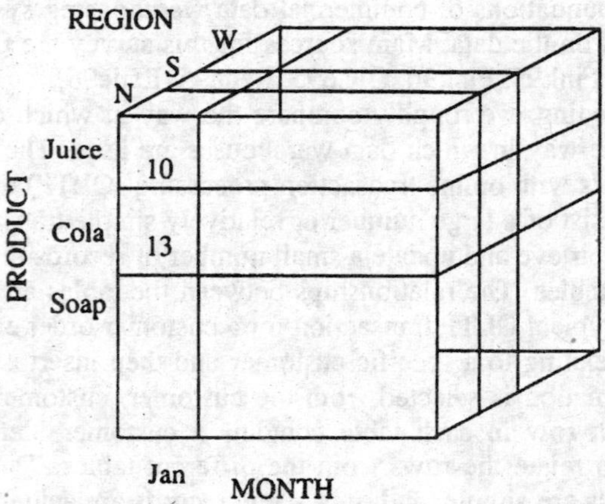

Fig. 5.1. Sales volume as a function of product, time, and geography

Measures (which are also known as variables or metrics) – like *Sales* in the example, or budget, revenue, inventory, etc. – in a multidimensional array correspond to columns in a relational database table whose values functionally depend on the values of other columns. Values within a table column correspond to values that measure in a multidimensional array: measures associate values with points in the multidimensional world. In our example a measure of the sales of the product *Cola*, in the northern region, in January, is 13. Thus, a dimension acts as an index for identifying values within a multidimensional array. If one member of the dimension is selected, then the remaining dimensions in which a range of members (or all members) are selected define a subcube. If all but two dimensions have a single member selected, the remaining two dimensions define a spreadsheet (or a "slice" or a "page"). If all dimensions have a single member selected, then a single cell is defined. Dimensions offer a very concise, intuitive way of organizing and selecting data for retrieval, exploration, and analysis.

Usual predefined dimension levels (or "roll-ups") for aggregating data in a DW are *temporal* (e.g., year vs. month), *geographical/spatial* (e.g., Rome vs. Italy), *organizational* (meaning the hierarchical breakdowns of an organization, e.g., institute vs. department), and *physical* (e.g., car vs. engine). Figure 5.2 shows some typical hierarchical summarization paths.

A value in a single cell may represent an aggregated measure computed from more specific data at some lower level of the same dimension. For example, the value 13 for the sales in January may have been consolidated as the sum of the disaggregated values of the weekly (or day-by-day) sales.

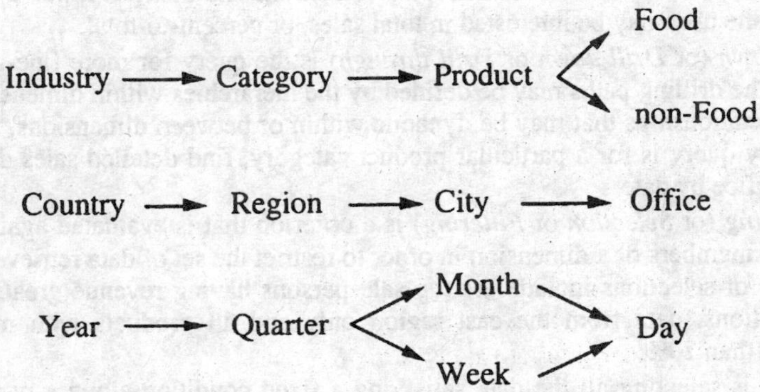

Fig. 5.2. Hierarchical summarization paths

The rest of this chapter is organized as follows. We first provide a general view of the area of practice, pointing out the different approaches that have been taken in the commercial systems for representing and handling multidimensional information. We also give a structured list of the most important tools available in the market. The research problems, which are highlighted by the practical approaches pursued by the commercial products, are then discussed. Finally, the last sections review the state of the art of research work that has been done in the data warehouse field, as well as of research work in knowledge representation and conceptual modeling closely related to the problems mentioned above.

5.1 Multidimensional View of Information

As we said before, the multidimensional view of data considers that information is stored in a multidimensional array or cube. A cube is a group of data cells arranged by the dimensions of the data, where a dimension is a list of members, all of which are of a similar type in the user's perception of the data. Each dimension has an associated hierarchy of levels of aggregated data, i.e., it can be viewed at different levels of detail (e.g., *Time* can be detailed as *Year, Month, Week,* or *Day*). *Navigation* (often improperly called *slicing and dicing*) is a term used to describe the processes employed by users to explore a cube interactively by drilling, rotating, and screening, generally using a graphical OLAP client connected to an OLAP server. The result of a multidimensional query is a cell, a two-dimensional slice, or a multidimensional subcube. The most popular end-user operations on multidimensional data in commercial systems are the following:

- *Aggregation* (or *Consolidate, Roll-up*) is the querying for summarized data. Aggregation involves two different tasks. First, the data relationships (according to the attribute hierarchy within dimensions or to cross-dimensional formulas) for the dimensions the user wants to see on a more coarse-grained level must be considered. Second, the new measure must be computed with respect to these more coarse-grained levels and the specified aggregation function. For example, sales offices can be rolled up to districts and districts rolled up to regions; the user may be interested in total sales, or percent-to-total.
- *Roll down* (or *Drill down* or *Drill through*) is the query for more fine-grained data. The drilling paths may be defined by the hierarchies within dimensions or other relationships that may be dynamic within or between dimensions. An exemplary query is for a particular product category, find detailed sales data for each office by date.
- *Screening* (or *Selection* or *Filtering*) is a criterion that is evaluated against the data or members of a dimension in order to restrict the set of data retrieved. Examples of selections include the top salespersons having revenue greater than 12 millions, data from the east region only and all products with margins greater than 20%.
- *Slicing* is selecting all the data satisfying a fixed condition along a particular dimension while navigating. A slice is a subset of a multidimensional array where a single value for one or more members of a dimension has been specified. For example, if the member *Actual* is selected from a *Scenario* dimension, then the subcube of all the remaining dimensions is the slice that is specified. The data omitted from this slice would be any data associated with the non-selected members of the Scenario dimension, e.g., Budget, Variance, Forecast, etc. From an end user perspective, the term slice most often refers to the visualization of a two-dimensional page (a spreadsheet) projected from the multidimensional cube.

Products	Q1	
Electronics	$5.2	
Toys	$1.9	
Clothing	$2.3	

Sales
Year to date ($ millions)

Electronics	Q1	Q2	Q3
VCR	$1.4		
Camcorder	$0.6		
TV	$2.0		
CD Player	$1.2		

Fig. 5.3. *Drill down* presents increasingly detailed levels of data

- *Scoping*. is restricting the view of database objects to a specified subset. Further operations, such as update or retrieve, will affect only the cells in the specified subset. For example, scoping allows users to retrieve or update only the sales data values for the first quarter in the east region, if these are the only data they wish to receive. While conceptually Scoping is very similar to Screening, operationally Scoping differs from Screening as being only a sort of preprocessing step in a navigation phase.
- *Pivot* (or *Rotate*) is to change the dimensional orientation of the cube, for analyzing the data using a particular dimension level as independent variable. For example, if we consider a two-dimensional array – i.e., a spreadsheet – rotating may consist of swapping the rows and columns of the spreadsheet itself, moving one of the row dimensions into the column dimension, or swapping an off-spreadsheet dimension with one of the dimensions in the page display (in order to become one of the new rows or columns), etc. A specific example of the first case would be taking a report that has Time across (the columns) and Products down (the rows) and rotating it into a report that has Product across and Time down. An example of the second case would be to change a report which has Measures and Products down and Time across into a report with Measures down and Time over Products across. An example of the third case would be taking a report that has Time across and Product down and changing it into a report that has Time across and Geography down.

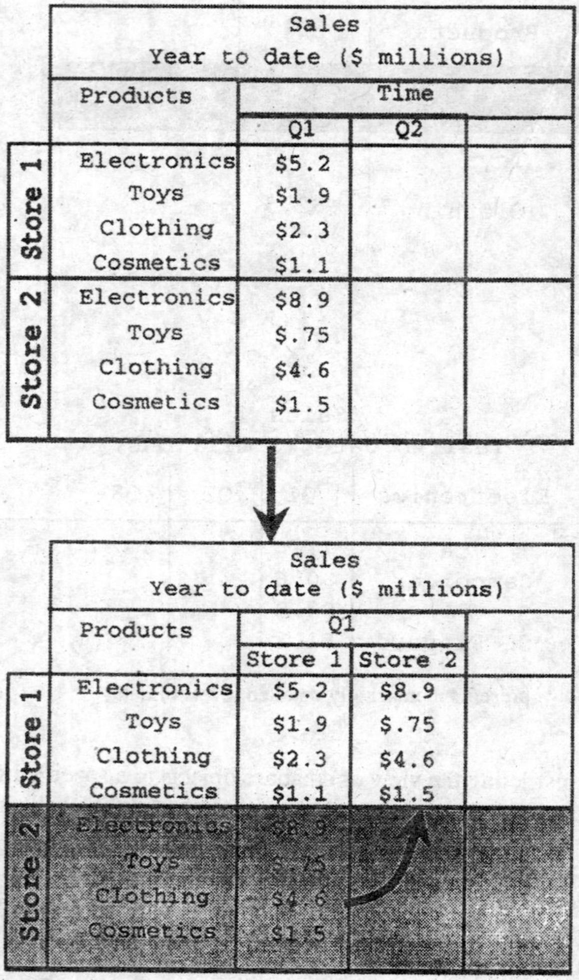

Fig. 5.4. *Pivoting*: here sales data moves side by side for easier comparison

5.2 ROLAP Data Model

Given the popularity of relational DBMS, one is tempted to build an OLAP tool as a semantic layer on top of a relational store [Coll96]. This is called *Relational OLAP* (*ROLAP*). The basic idea underlying this approach is to use an extended relational data model according to which operations on multidimensional data are mapped to standard relational operations. This layer provides a multidimensional view, computation of consolidated data, drill-down operations, and generation of appropriate SQL queries to access the relational data.

Commercially available ROLAP tools usually offer a multiplicity of operations: they have a SQL generator, capable of creating multipass selects and/or correlated subqueries; they are powerful enough to create nontrivial ranking, comparison and

%-to-class calculations; they try to generate SQL optimized for the target database, including SQL extensions, for example, using *Group By, Correlated* subquery, *Having, Create View*, and *Union* statements; they provide a mechanism to describe the model through metadata, and use the metadata in real time to construct queries; they may include a mechanism to at least advise on the construction of summary tables for performance. However, data warehouse users report that often the SQL code optimizers are still generating rather inefficient queries which can cause serious performance problems.

At the heart of the relational OLAP technologies is dimensional modeling. This technique organizes information into two types of data structures: measures, or numerical data (e.g., sales and gross margins), which are stored in "fact" tables; and dimensions (e.g., fiscal quarter, account, and product category), which are stored in satellite tables and are joined to the fact tables. ROLAP systems must provide technologies capable of optimizing the following three database functions:

- *Denormalization*, a database design that repetitively stores data in tables, minimizing the number of time-consuming joins when executing a query, and reducing the number of rows that must be analyzed;
- *Summarization*, a technique for aggregating (consolidating) information in advance, eliminating the need to do so at run time;
- *Partitioning*, the ability to divide a single large fact table into many smaller tables, thereby improving response time for queries as well as for data warehouse backup and reloading.

An argument against the use of a ROLAP model is about poor performance due to multiple joins [Fink96]. Let us consider an example. A *Sales* database might contain the following tables and data elements:

Table1	Product Sales/Sales Office	1,000,000 rows
Table2	Product Description	1,000 rows
Table3	Sales Office/District Cross Ref	100 rows
Table4	District/Region Cross Ref	10 rows

The above normalized database saves space because each product, sales office, district office, and region appears only once in the database. Database designers decompose related data into normalized relational tables to èliminate redundant data, which is difficult to maintain and can lead to inconsistent data updates. Redundant data can also greatly increase disk space requirements. However, what is good for update-oriented (such as OLTP) applications is not necessarily good for analytical applications. In fact, a query that needs to summarize and compare data by sales office, district, and regions can be very expensive since it has to join the four tables together. In this database, the join might require up to one trillion matches (1 000 000 * 1 000 * 100 * 10). A database consisting of ten or more tables would take several times more than this. In the worst case, this matching process must be performed for every OLAP query.

To overcome this problem, *data marts* are introduced in ROLAP. Data marts are collections, possibly in the form of materialized views derived from a source data warehouse, where aggregations and partitions are implemented according to the needs of targeted decision makers, so that queries can be better optimized.

They are subdivided into "standard collections" and "consumer collections," depending on the user analysis model.

Data marts are special denormalized databases: denormalized tables are tables that are prejoined (i.e., all tables are combined into one table) to avoid time consuming joins. Unfortunately, there are several disadvantages to this approach: basically, denormalization is expensive in terms of both performance and system resources. Denormalization produces extremely large tables because data are stored redundantly in each record. Online analytical queries are then forced to scan very large tables resulting in queries which can be just as expensive, if not more expensive, than table joins. Moreover, denormalization increases the sparseness (emptiness) in a database. Suppose a product sales record can also be associated with a warehouse and 25% of the product sales records have a warehouse associated with them and 75% do not. When the denormalized table is created, 75% of the records will not contain any information in the warehouse field. So, while denormalization eliminates joins, it does not necessarily increase performance. In fact, performance can be much worse depending upon the query.

The main approach followed in practice to overcome these usability and performance problems within the relational model is by storing data in a "star" structure, which tries to superimpose a multidimensional structure upon the two-dimensional relational model. The star model simplifies the logical model by organizing data more efficiently for analytical processing.

A star schema consists of one central fact table and several dimension tables. The measures of interest for OLAP are stored in the fact table (e.g., sales, inventory). For each dimension of the multidimensional model there exists a dimensional table (e.g., region, product, time). This table stores the information about the dimensions.

Figure 5.5 is an example of a star schema. The table *Sales* at the center of the star is the fact table with the foreign keys *geography code*, *time code*, *account code*, and *product code* to the corresponding dimension tables, which carry the in

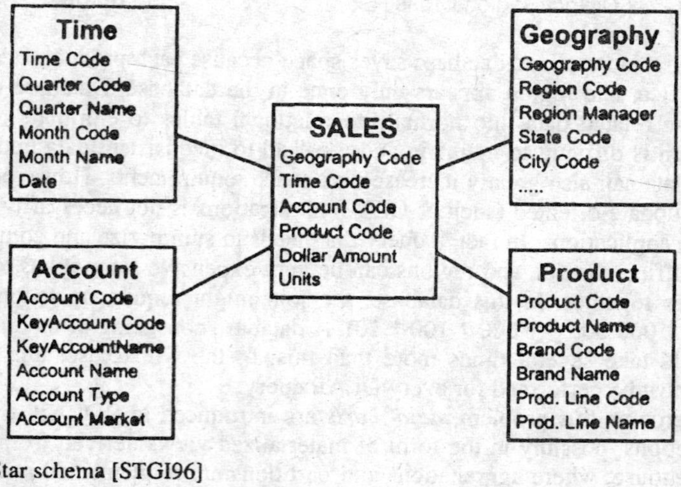

Fig. 5.5. Star schema [STGI96]

formation on the hierarchical structure of the respective dimensions. Each dimension table consists of several attributes describing one dimension level. For example, the product dimension is organized in a *product, product line,* and *brand* hierarchy. The hierarchical structure of a dimension can then be used for drill-down and roll-up operations. Every hierarchy is represented by one or more attributes in the dimension table. Dimension tables have a denormalized structure. Because of the aforementioned problems with denormalized data structures, such as redundancy and waste of storage, it might be useful to create a *snowflake* schema. This is done by normalizing the dimension tables of the star schema. A snowflake schema corresponding to the previous star schema is illustrated in Fig. 5.6. For example, the time table is now split into three new tables: *Time, Quarter,* and *Month.*

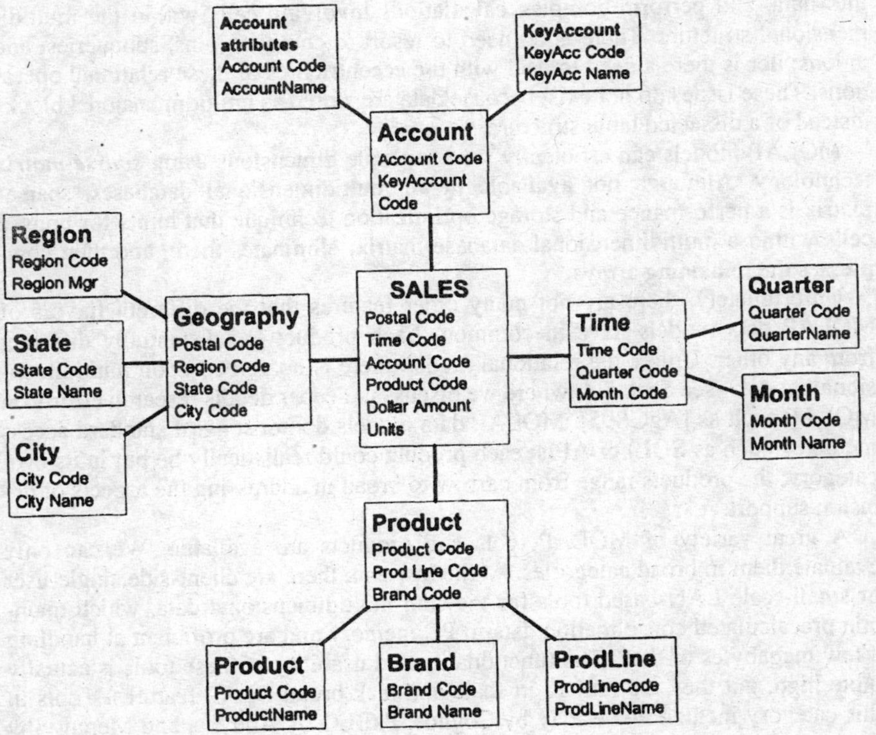

Fig. 5.6. Snowflake schema [STGI96]

5.3 MOLAP Data Model

In contrast to ROLAP, *Multidimensional OLAP (MOLAP)* is a special purpose data model in which multidimensional data and operations are mapped directly.

Rather than storing information as records, and storing records in tables, *Multidimensional Databases (MDDs)* store data in arrays. Multidimensional databases are capable of providing very high query performance, which is mostly accomplished by anticipating and restricting the manner in which data will be accessed. In general, information in a multidimensional database is of coarser granularity than the one in a standard relational database; hence, the index is smaller and is usually resident in memory. Once the in-memory index is scanned, a few pages are drawn from the database. Some tools are designed to cache these pages in shared memory, further enhancing performance. Another interesting aspect of multidimensional databases is that information is *physically* stored in arrays: this means that values in the arrays can be updated without affecting the index. A drawback of this "positional" architecture is that even minor changes in the dimensional structure require a complete reorganization of the database.

Another advantage of MOLAP with respect to ROLAP is the possibility of having *native* OLAP operations. With an MDD server, it is easy to pivot information (since all the information is stored in one multidimensional hypercube), drill down into data, and perform complex calculations involving cells within the multidimensional structure. There is no need to resort to complex joins, subqueries, and unions; nor is there a need to deal with the eccentricities of these relational operations. These issues do not exist because data are stored as multidimensional blocks instead of a dissected table structure.

MOLAP models can efficiently store multiple dimensions using *sparse-matrix* technology. Although not available in all multidimensional databases, sparse-matrix is a performance and storage optimization technique that hunts for unused cells within a multidimensional database matrix, eliminates them, and then compresses the remaining arrays.

Unfortunately, there are not many other features that the different flavors of MOLAP data models have in common. Each product is substantially different from any other. Unlike the relational model, there is no agreed-upon multidimensional model – see Sect. 5.4, where we discuss in deeper details research efforts on MOLAP such as [AgGS95]. MOLAP data models do not suggest standard access methods (such as SQL) or APIs; each product could realistically be put in its own category; the products range from narrow to broad in addressing the aspects of decision support.

A great variety of MOLAP tools and products are available. We can only evaluate them in broad categories. At the low end, there are client-side single-user or small-scale LAN-based tools for viewing multidimensional data, which maintain precalculated consolidation data in PC memory and are proficient at handling a few megabytes of data. The functionality and usability of these tools is actually quite high, but they are limited in scale and lack broad OLAP features. Tools in this category include PowerPlay by Cognos, PaBLO by Andyne, and Mercury by Business Objects; all of them implement an underlying ROLAP data model. At the high end, tools like Acumate ES by Kenan, Express by Oracle (both explicitly implementing a MOLAP data model), Gentium by Planning Sciences, and Holos by Holistic Systems are so broad in their functionality that each of these tools could realistically define a separate category, so diverse are their features and architectures. The pure MDDB engines supporting the MOLAP data model are represented by Essbase by Arbor, LightShip Server by D&B/Pilot, MatPlan by Thinking Networks, and TM/1 by Sinper.

5.4 Logical Models for Multidimensional Information

We can divide the research attempts and problems regarding multidimensional aggregation into three major categories: *conceptual*, *logical* and *physical* fields. In the conceptual field, the major issue is to develop new or extend standard conceptual modeling formalisms (e.g., the entity-relationship (ER) diagrams) to cope with aggregation. In the logical field, the research focuses on the development of a logical, implementation independent model, to describe data cubes. Finally, the problems of physical storage, update propagation, and query evaluation belong to the physical field. In this chapter, we will discuss only logical and conceptual models. As for the physical field, we will deal with the optimal data warehouse design problem in Sect. 7.5.2, and with the indexing and query evaluation issues in Chap. 6, dealing with query optimization.

Following [BSHD98], a *logical* model for data cubes must satisfy the properties listed below:

- Independence from physical structures
- Separation of structure and content
- Declarative query language (i.e., a calculus-like language)
- Complex, structured dimensions
- Complex, structured measures

One could add more requirements to this list, such as:

- Power to deal with sequences of operations (since this is actually what a system would perform in practice)
- Completeness of operations (i.e., a set of algebraic operations powerful enough to capture all the usual operations performed by an OLAP system)

In the following, we give a brief survey of the most important and influential logical models of MDDs.

Gray et al. [GBLP96] expand a relational table by computing the aggregations over all the possible subspaces created from the combinations of the attributes of such a relation. Practically, the CUBE operator that is introduced calculates all the marginal aggregations of the detailed data set. The OLAP extensions to the Microsoft SQL Server are based on the *data cube* operator introduced in [GBLP96].

In [AgGS95], a model for MDDs is introduced (Fig. 5.8). The model is characterized by symmetric treatment of dimensions and measures. A minimal set of operators deals with the construction and destruction of cubes, join and restriction of cubes and merging of cubes through direct dimensions. Furthermore, a translation of these cube operators into SQL queries is given.

Li and Wang [LiWa96] introduce a multidimensional data model based on relational elements. Dimensions are modeled as "dimension relations," practically annotating attributes with dimension names. A single cube is modeled as a function from the Cartesian product of the dimensions to the measure and can be mapped to "grouping relations" through an applicability definition. A grouping algebra is presented, extending existing relational operators and introducing new ones, such as ordering and grouping to prepare cubes for aggregations. Furthermore, a multidimensional algebra is presented, dealing with the construction and modification

```
SELECT Model, Year, Color, SUM(sales)
       AS Sales
FROM Sales
WHERE Model in ('Ford', 'Chevy')
  AND Year BETWEEN 1990 AND 1992
GROUP BY Model, Year, Color WITH CUBE;
```

SALES

Model	Year	Color	Sales
Chevy	1990	red	5
Chevy	1990	white	87
Chevy	1990	blue	62
Chevy	1991	red	54
Chevy	1991	white	95
Chevy	1991	blue	49
Chevy	1992	red	31
Chevy	1992	white	54
Chevy	1992	blue	71
Ford	1990	red	64
Ford	1990	white	62
Ford	1990	blue	63
Ford	1991	red	52
Ford	1991	white	9
Ford	1991	blue	55
Ford	1992	red	27
Ford	1992	white	62
Ford	1992	blue	39

CUBE

DATA CUBE

Model	Year	Color	Sales
Chevy	1990	blue	62
Chevy	1990	red	5
Chevy	1990	white	95
Chevy	1990	ALL	154
Chevy	1991	blue	49
Chevy	1991	red	54
Chevy	1991	white	95
Chevy	1991	ALL	198
Chevy	1992	blue	71
Chevy	1992	red	31
Chevy	1992	white	54
Chevy	1992	ALL	156
Chevy	ALL	blue	182
Chevy	ALL	red	90
Chevy	ALL	white	236
Chevy	ALL	ALL	508
Ford	1990	blue	63
Ford	1990	red	64
Ford	1990	white	62
Ford	1990	ALL	189
Ford	1991	blue	55
Ford	1991	red	52
Ford	1991	white	9
Ford	1991	ALL	116
Ford	1992	blue	39
Ford	1992	red	27
Ford	1992	white	62
Ford	1992	ALL	128
Ford	ALL	blue	157
Ford	ALL	red	143
Ford	ALL	white	133
Ford	ALL	ALL	433
ALL	1990	blue	125
ALL	1990	red	69
ALL	1990	white	149
ALL	1990	ALL	343
ALL	1991	blue	106
ALL	1991	red	104
ALL	1991	white	110
ALL	1991	ALL	314
ALL	1992	blue	110
ALL	1992	red	58
ALL	1992	white	116
ALL	1992	ALL	284
ALL	ALL	blue	339
ALL	ALL	red	233
ALL	ALL	white	369
ALL	ALL	ALL	941

Fig. 5.7. The CUBE operator [GBLP96]

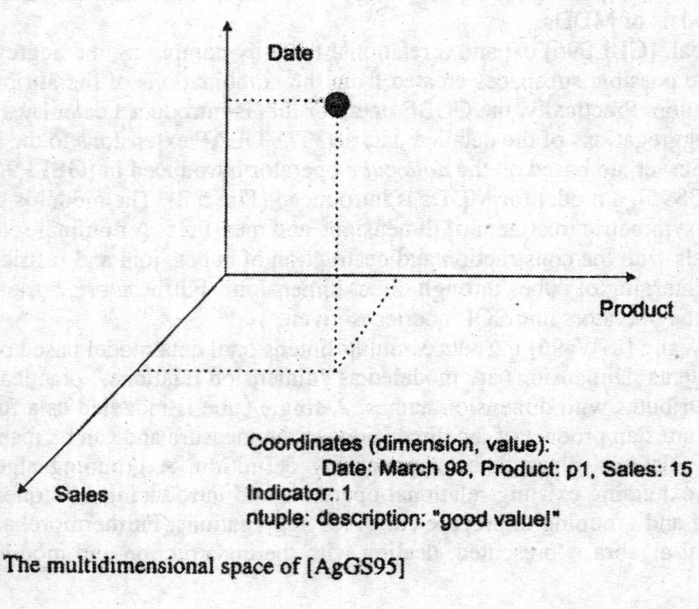

Fig. 5.8. The multidimensional space of [AgGS95]

relation g

A1	A2	A3
a	4	c
a	4	h
b	2	b
c	1	d

order on A1

A1
a
b
c

$roll^{1,1,2}(g)$

s-A1	end -A1	A1	A2	A3
a	b	a	4	c
a	b	a	4	h
a	b	b	2	b
b	c	b	2	b
b	c	c	1	d

Fig. 5.9. Operator roll [LiWa96]

of cubes as well as with aggregations and joins. A relation can be grouped by intervals of values. The values of the "dimensions" are ordered and then grouped using an auxiliary table.

Gyssens and Lakshmanan [GyLa97] define *n*-dimensional tables and a relational mapping based on the notion of *completion*. An algebra (and an equivalent calculus) is defined with classical relational operators as well as restructuring, classification, and summarization operators. The algebra is expressive enough to capture the notion of a data cube and to represent monotone roll-up operators.

In [CaTo97], a model for MDDs is presented that is based upon the notions of dimensions and *f-tables*. Dimensions are constructed from hierarchies of dimension levels, whereas f-tables are repositories for the factual data. A cube is then characterized by a set of *roll-up* functions, mapping the instances of a dimension level to instances of another dimension level. A query language is the focus of this work: a calculus for f-tables along with scalar and aggregate functions is presented, basically oriented towards the formulation of aggregate queries. In [CaTo98], the focus is on the modeling of multidimensional databases: the basic model remains practically the same, whereas ER modeling techniques are given for the conceptual modeling of the multidimensional database. In [CaTo98a], a graphical query language as well as an equivalent algebra is presented. The algebra is a small extension to the relational algebra, including a roll-up operator, yet not equivalence to the calculus is provided.

In [Vass98], dimensions and dimension hierarchies are explicitly modeled. Furthermore, an algebra representing the most common OLAP operations is provided. The model is based on the concept of *basic cube* representing the cube with the most detailed information (i.e., the information with respect to the lowest levels of the dimension hierarchies). All other cubes are calculated from the basic cubes using terms the algebra. The algebra allows the execution of sequences of operations as well as for drill-down operations. A relational mapping is also provided for the model, as well as a mapping to multidimensional arrays.

In [Lehn98], a model for multidimensional databases is presented, based on primary and secondary multidimensional objects. A *Primary Multidimensional*

Object (PMO), which represents a cube, consists of a cell identifier, a schema definition, a set of selections, an aggregation type (sum, avg, no-operator), and a result type. A *Secondary Multidimensional Object* (SMO) consists of all the dimension levels (also called "dimensional attributes") to which one can roll up or drill down for a specific schema. Operations like roll up, drill down, slice, dice, etc. are also presented. In [LeAW98], which is a continuation of the previous paper, two multidimensional normal forms are proposed, defining (a) modeling constraints for summary attributes and (b) constraints to model complex dimensional structures.

A recent rediscovery in the research on multidimensional data models is their striking similarity with the multidimensional data models studied in the Statistical Databases community. In [Shos97], a comparison of work done in statistical and multidimensional databases is presented. The comparison is made with respect to application areas, conceptual modeling, data structure representation, operations, physical organization aspects, and privacy issues. The basic conclusion is that the two areas have a lot of overlap, with statistical databases emphasizing conceptual modeling and OLAP emphasizing physical organization and efficient access.

In [OzOM85], a data model for statistical databases is introduced. The model defines operators on "summary tables" such as construction/ destruction, concatenation/ extraction, attribute splitting/ merging, and aggregation operators. The underlying algebra is a subset of the algebra described in [OzOM87]. Furthermore, physical organization and implementation issues are discussed. [OzOM85] is very close to practical OLAP operations, although discussed in the context of summary tables.

In [RaRi91], a functional model ("*Mefisto*") is presented. Mefisto is based on the definition of a data structure, called a "statistical entity," and on operations defined on it like summarization, classification, restriction and enlargement.

5.5 Conceptual Models for Multidimensional Information

Due to the presence of multidimensional aggregation, data warehouse – and especially OLAP – applications ask for the vital extension of the expressive power and functionality of traditional conceptual modeling formalisms. Still, there have been few attempts [AgGS95, DeLe95] to provide such an extended modeling formalism, despite the fact that (1) experiences in the field of databases have proven that conceptual modeling is crucial for the design, evolution, and optimization of a database; (2) a great variety of data warehouse systems are on the market, most of them providing some implementation of multidimensional aggregation; and (3) query optimization is even more crucial for data warehouses than it is for databases – which makes semantic query optimization using a conceptual model even more important. It turns out, however, that a comparison of different systems or language extensions for query optimization is difficult: a common framework in which to translate and compare these extensions is missing. New query optimization techniques developed for extended schema and/or query languages (see [GuHQ95, LeMu96, SDJL96, MuSh95] for query optimization with aggregation, and [LeRO96] for planning queries to heterogeneous sources) cannot be compared appropriately. In most cases, it can be easily seen that the optimization algorithms transform queries to equivalent queries, but it remains open where one algorithm is better than another one, whether it is optimal or in how far it is incomplete.

Summarizing, a formal framework must be developed that encompasses the data warehouse related extensions of traditional representation formalisms. We refer to this formalism by *Conceptual Data Warehouse Formalism* (CDWF). In order to overcome the above mentioned lack of formalization of semantic data warehouse problems, a CDWF

- Has to be equipped with well-defined semantics
- Should be expressive enough to capture the data models relevant in data warehouse applications
- Should give a formalization of the operators on the data structures used in data warehouse applications
- Has to be able to capture inference problems relevant for reasoning in data warehouses like query optimization

A CDWF satisfying these properties will be an important step towards understanding and comparing different data warehouse systems, and hence for the evaluation of their quality.

A CDWF provides the language for defining *multidimensional* information within a conceptual schema in the data warehouse information base. As stated above, the schema is of support for the conceptual design of a data warehouse, for query and view management, and for update propagation: it serves as a *reference meta schema* for deriving the interrelations among entities, relations, aggregations, and for providing the integrity constraints necessary to reduce the design and maintenance costs of the data warehouse. A CDWF must be expressive enough to describe both the abstract business domain concerned with the specific application (enterprise model) – just like a conceptual schema in the traditional database world – and the possible views of the enterprise information a user may want to analyze (client model) – with particular emphasis on the aggregated views, which are peculiar to a data warehouse architecture. A multidimensional modeling object in the logical perspective – e.g., a materialized view, a query, or a cube – should always be related with some (possibly aggregated) entity in the conceptual schema.

According to a conservative point of view, a desirable CDWF should extend some standard modeling formalism (such as Entity-Relationship or OMT) to allow the description of both aggregated entities of the domain – together with their properties and their relationships with the other relevant entities – and the dimensions involved.

A very promising proposal has been done in the context of statistical data modeling by [DeNa96] and [CaDL95]. The authors propose a CDWF that satisfies most of the above mentioned properties – in particular, it has a powerful structuring mechanism – and for which the interesting specialized reasoning services described in the next section have been defined.

5.5.1 Inference Problems for Multidimensional Conceptual Modeling

Assuming that the syntax, semantics, and operators of the CDWF are defined according to the requirements stated above, it remains to specify relevant inference problems, to investigate these problems with respect to their computational complexity, and to develop reasoning algorithms for them. This will provide a theoretical and algorithmic basis which can be used for the design and evolution of a data warehouse and for semantic query optimization.

As in traditional representation formalisms, many inference problems can be reduced to satisfiability and containment. Satisfiability of classes (or of queries, by means of the classes they represent) is the problem if there exists a world such that each class of a given set of classes has at least one instance. Finally, containment asks if one class is more general than another one, that is, whether each instance of the latter class is always also an instance of the more general one. It is well known that solutions to these problems can be used for optimizing queries: for example, if a query (respectively the class it represents) is not satisfiable, we do not need to process it since its result is surely empty. If a query is contained in a materialized view, then this view can be used to process the query instead of searching in a larger table.

Now, data warehouse applications confront us with a third inference problem. Aggregation is the central means to summarize and condense the information contained in the various sources. It occurs (1) when integrating data from sources, (2) when building views for the data marts, and (3) in ad-hoc queries. As queries to the sources or to larger views are far more expensive than those to smaller views, we are confronted with a new problem, namely, given a query involving aggregation and a (materialized) view, can this query be computed using (the aggregations contained in) this view? This depends on whether the aggregations contained in the view are still "fine-grained enough" to compute the aggregations required by the query. For example, suppose that a user asks the system to compute a query Q, namely the total profit of all product groups for each year and each region. If a (materialized) view V exists which contains the profit for the product groups food and nonfood products for all quarters for all regions, then the total profit can be computed by simply summing up those partial profits for each year (given that we sell only food and nonfood and that nothing is both food and nonfood). Please note that the query Q is not contained in the view V in the classical reading of containment. Nevertheless, V can be used to compute Q.

In the following, we will call this relationship between a query (or a view) and a view (or a query) refinement. The main difference between the containment relation and the refinement relation is the following: for a view V' to be contained in another view V, each element of V' is contained in V, and, roughly spoken, it can be obtained simply by erasing some lines or columns of V. For a view V' to be more coarse-grained than another view V, erasing is no longer sufficient. It might be necessary to aggregate some elements of V to build an element of V'.

As the last reasoning task to be cited here, there is the retrieval of all those instances in a given database which satisfy certain properties. Traditionally, these properties are specified using an expressive query language like SQL, conjunctive queries, QBE, etc. This high expressiveness is possible since, in general, the answer to a query (that is, to retrieve all instances satisfying the query) is less complex than deciding, for example, if a given query is satisfiable.

Summing up, we are confronted with four reasoning services or problems:

- To decide whether queries (or views) are satisfiable,
- Whether one is contained in another,

- Whether one is refined by another,
- To answer a query.

The first three reasoning problems belong to the set of intentional reasoning problems, whereas the last one belongs to the set of extensional reasoning problems. In general, intentional reasoning is more complicated than extensional reasoning. As a consequence, given that all these problems should be decidable, one may use a more expressive language to formulate extensional problems than the language used to formulate intentional problems.

A logical approach for reasoning is surely useful not only for integrating heterogeneous sources and optimizing (aggregate) queries but also for update propagation, data warehouse design and evolution. In update propagation, the information provided by integrity constraints (expressed in some logic) can be used to reduce the maintenance cost of the data warehouse. This information along with reasoning mechanisms for checking query containment or query refinement (more generally, query rewriting over views) can be used for optimal and incremental data warehouse design, as well as data warehouse evolution.

5.5.2 Which Formal Framework to Choose?

As a starting point for the development of CDWF, we propose *description logics* (DL) for the following reasons. First, description logics can be seen as a unified framework for class-based representation formalisms [CaLN94, DeLe95, Borg94, BJNS94], including powerful extensions of the entity-relationship model. Second, several extensions of description logics by concrete domains (built-ins) [BaHa91, BaHa93, KaWa96] have already been investigated. Third, powerful reasoning algorithms have been designed [DeDM96, DeLe96]. The relevance of reasoning techniques developed in description logics for data warehouse applications has been stressed in [GBBI96, Rudl96]. And finally, description logics are equipped with well-understood formal semantics. In fact, description logics satisfy all the requirements specified for a CDWF at the beginning of this section.

However, to serve as a CDWF, the expressive power of description logics must be further increased so that multidimensional aggregation can be represented adequately. More precisely, the target description logic must be able to represent aggregated objects, hierarchically structured dimensions (such as Food, Products, Time, etc.), and aggregation functions on concrete domains (such as sum, min, average, etc., on the reals, integers, etc.). Concerning the hierarchically structured dimensions, the description logic formalism should provide predefined dimensions (such as time or space) as well as means to build user-defined hierarchically structured dimensions. In the following, these points will be explained in more detail.

Hierarchically structured dimensions: In order to support multiple hierarchies, the description logics must provide means for defining these hierarchies, and the query language should allow for arbitrary aggregation along the hierarchies. A first approach is to provide means for the definition of partitions. An example of a multiple hierarchy of product groups is given to illustrate this idea:

```
(def-partition use          :divides products
    :into (food, nonfood))
(def-partition origin       :divides products
    :into (imported, self-made))
(def-partition state        :divides food
    :into (raw, prepared))
(def-partition clients      :divides food
    :into (veget., lacto-veget., meat-cont.)
(def-partition staple       :divides raw
    :into (vegetable, fruit, grain, meat, seafood))
```

Note that `products` is not meant to be a set of products, but it denotes individual product groups decomposed into, for example, the group `food` and the group `nonfood`. Following this approach, attributes (like `total-sales`) still relate individual objects to other individual objects. This would not be the case if `products` were the *set* containing all single products: in this case, the attribute `total-sales` would relate a set to an element of, say, the integer domain.

Such a hierarchy can then be used to define new, complex aggregates. For example, one can describe all those sales relating to the same product with respect to the level `staple`, namely `vegetables`, `fruits`, etc.

Regarding dimensions, the following important questions have also to be addressed:

- The partitions like `use`, `origin`, `state` can be viewed as particular part/whole relations. In how far does this view yield stronger consequences? For example, in how far can the transitivity of the general part-whole relation be exploited in the reasoning process? Are there certain features (like `sales` in the example), whose value is always equal to the sum of the respective values of its parts? For a survey of part/whole relations in Knowledge Representation see [AFGP96].
- The inclusion of a specific treatment for the temporal and spatial dimensions at the conceptual level. How difficult is to reason with an extension of the CDWF with an explicit representation of time and space? For a survey on temporal extensions of description logics see [ArFr98].

Aggregation functions: In [BaHa91], description logics are extended with concrete domains consisting of a domain (like the real numbers) and built-in predicates over this domain (like `less-than`, `divides`, etc.). Inference algorithms for intentional reasoning services for this hybrid description logics ALC(D) are presented which are provably sound and complete, provided that the concrete domain satisfies a rather weak condition. This approach can be extended by allowing the concrete domain to carry a set of aggregation functions (like sum, average, etc.) which are used to define new attributes. For example, given the above partition of the product palette, the total sales of all products can be described by `sum(use ∘ sales)`. An important recent result identifies the borders for the possible extensions of a CDWF towards the inclusion of multidimensional aggregation [BaSa98]. It has turned out that the explicit presence of arbitrary aggregation functions, when viewed as a means to define new attribute values for aggregated entities, and built-in predicates in a concrete domain, increases the expressive power of the basic description logic in such a way that all interesting inference problems may easily become undecidable. In fact, a very general result of undecidability with simple grouping and aggregation functions (sum, min, max) over any domain, including the natural numbers, has been proved. There are two ways which can be pursued in order to gain decidability : by restricting the concrete domain (e.g., to min, max), or by restricting the underlying abstract part of the description logic.

5.6 Conclusion

As shown in this chapter, a theory of multidimensional data models – both at the logical and conceptual models – is only slowly emerging from a confluence of results stemming from practice, statistical database theory, extensions of relational theory, and logic-based conceptual modeling.

The practical value of such studies lies in the definition and efficient execution of user queries, which is the subject of the following chapter. Part of this perspective is the question of multidimensional visualization. An in-depth discussion is beyond the scope of this book.

In a longer term, such data models will also provide better foundations for optimal data warehouse design and evolution, as discussed in the final chapter of this book.

6 Query Processing and Optimization

The ultimate purpose of a data warehouse is to support queries by end users who want to analyze the available information for an organization. However, from a more abstract point of view, queries are not only processed at the data warehouse back end.

As pointed out earlier, a data warehouse can be seen as a collection of materialized views that result from querying the underlying sources. Source-view relationships are also present within a data warehouse. In general, data warehouses are hierarchically structured: there are different layers of views, where the views in one layer are derived from those in the layers beneath. As we will see, the views in a data warehouse share many characteristics with user queries and computing them poses similar problems. For this reason, we will discuss query processing in a setting as general as possible, which comprises all parts of the data warehouse architecture.

A data warehouse not only contains "object data" describing a company or other organization but also auxiliary data – usually called "metadata" – describing the internal structure and the operations of the data warehouse. Browsing and querying metadata is an important part of the activities of each end user, who has to know several characteristics of information (e.g., structure, granularity) before querying for it. Moreover, metadata are needed by administrators to monitor the data warehouse. Finally, the operations of the system itself are specified and controlled through metadata.

However, in contrast to object data, there is no consensus for a standard formalism on how to represent meta information. In particular, there is no agreement on the format and content of metadata queries. We therefore restrict ourselves in this chapter to object-level queries.

The remainder of the chapter is organized as follows. We first give an overview of the state of practice; we identify which kind of queries occur in a data warehouse architecture environment and then we review in more detail the requirements on query processing in data warehouses and the business solutions currently offered. Next, we review the state of the art on the optimization of data warehouse queries. Finally, we discuss open problems.

6.1 Description and Requirements for Data Warehouse Queries

We discuss the setting in which queries occur in a data warehouse. Here, we understand queries in the general sense defined before. We suppose a simplified architecture, consisting of the *back end*, the *front end*, and the *core*.

Then, we describe the requirements on query processing in the different parts of a data warehouse and how they are met in practice. We consider first those characteristics of data warehouse queries that can be described in general terms, without referring to the particular data model according to which the data is organized. In particular, we discuss what distinguishes data warehouse queries from queries over transactional systems. Such a top-level point of view is sufficient to understand properties of the queries in the back end and the core.

6.1.1 Queries at the Back End

The *back end* connects the data warehouse with the operational data sources. With respect to the terminology introduced in Chap. 1, the back end consists of the sources, the Operational Data Store (ODS) and a set of intermediate wrappers for the transportation, cleaning and transformation of the data. We assume that each source is "wrapped" into a relational. That is, we assume there is an interface that lets each source appear as a relation with a finite set of attributes. When accessed, a source can output its content, either entirely or tuple by tuple or it can output those tuples which satisfy certain conditions imposed on their attributes.

The back end accesses the sources, usually at regular intervals, to load the data warehouse. Loading is more than producing a simple mirror of the sources.

- Loading involves "*data cleaning*", as already presented in Chap. 3. In the simplest case, cleaning is nothing but mapping cryptic codes occurring in the sources to more readable strings. More elaborate techniques, however, may involve accessing databases that contain extra information needed to resolve ambiguities in the sources.
- Loading may also involve *computing aggregates* of the source data, like sums, and other more or less sophisticated averages.

A more detailed description of the loading and the refreshment processes is found in Chap. 4, where we also elaborate on the issues of cleaning and aggregate view maintenance.

6.1.2 Queries at the Front End

The *front end* comprises the tools by which end users access the data warehouse. With respect to the terminology of Chap. 1, the front end consists of the OLAP/DSS tools and possibly their underlying data marts. There is a wide range of ways in which users retrieve information: from managers who have a quick look at a graphical user interface, to their assistants who inspect the latest version of a complex report, to analysts who query a data cube with a special tool for ad-hoc queries, to system personnel who build such interfaces and define the perspective under which the data warehouse can be perceived through the query tools.

6.1.3 Queries in the Core

The *core* consists of the data store for object and metadata, i.e., the primary data warehouse and the metadata repository. While schemata of OLTP systems are laid out in a way that keeps *redundancy* minimal, data warehouses, on the contrary, are deliberately designed in such a manner that the redundancy is inherent in the overall architecture: part of the data are views on other data, and thus redundant from a logical point of view. It can also be local to a component where precomputed results are held to speed up the response time for a class of queries that are expected to be posed frequently.

In corporate data warehouses, one often finds a *layered architecture* that reflects the structure of the organization. The bottom layer consists of regional data warehouses, each of which collects data from OLTP sources in a particular region. The regional data warehouses feed a central data warehouse that stores company-wide data. On top of the central data warehouse, there are data marts that present selected portions of the company-wide data warehouse to analysts. The data in the central data warehouse and in the data marts are views on the regional data warehouses. Thus, when the bottom layer in such an architecture is updated the views in the higher layers are recomputed to reflect the update.

In a data warehouse environment, users run decision support queries that require aggregation over large sets of data. The most common technique to speed up the execution of such queries is to *precompute aggregates* and store them together with the base data from which they are derived.

So far, we have identified view/base data relationships in the static structure of the data warehouse. A similar situation shows up when we observe the data warehouse over time. Data warehouses contain data at different levels of detail. The most recent data are stored at the greatest level of detail, while older data are condensed through aggregation. As time passes, data are moved from the detailed section for most recent data to the aggregated section for older data.

Redundancy increases the cost of updating. When base data are added through updates from the sources, the views derived from them have to be – partially or completely – reevaluated. A view can be recomputed either by executing again the query that defines it over the complete base data or by adopting view maintenance techniques, which derive the changes in the view from the changes in the base data and from some auxiliary information [GuMu95].

6.1.4 Transactional Versus Data Warehouse Queries

Queries over data warehouses are distinguished from queries over OLTP systems by their frequency and their volume. OLTP queries are parts of transactions. They touch only a few tuples but occur frequently (e.g., 50 transactions per second are possible in an airline's database). In contrast, data warehouse queries at all levels may touch up to several gigabytes of data but are issued at a much lower frequency. At the back end, 10, 000 queries per day posed by 100 users are about the maximal load that today's data warehouses can cope with [Kimb96]. Loading data from sources and materializing views is done far less often, usually at a daily or weekly rhythm.

It is a characteristic of data warehouse queries that they process huge sets of data. This does not mean, however, that they also produce voluminous output. Large sets of results are typical for queries that correspond to loading processes. In this respect, such queries resemble batch jobs running on a transactional database, like the monthly processing of a payroll.

Human users, in contrast, are not interested in gigantic output. They want to reduce large data volumes to a few characteristic parameters. To this purpose they need aggregation. Or they want to find a few exceptional data: "the needle in the haystack." Again, this is also achieved through a strategy that first computes statistical parameters and then identifies the deviant cases. Thus, as a rule, the closer a query is to the end user the more it requires aggregation.

Since data warehouse queries require the processing of large volumes of data, they tend to be time consuming. This makes query optimization a necessity. Moreover, there is a chance that more can be gained from optimizing complex voluminous queries (as in data warehouses) than from optimizing quick and simple queries. In particular, queries involving aggregations have a potential for optimization by applying selections early and thus reducing the set of tuples to be aggregated. Of course, such optimizations are most effective if they are supported by the right infrastructure. As a simple example, selection on an attribute can be performed most efficiently if there is an index for this attribute.

6.1.5 Canned Queries Versus Ad-hoc Queries

In a data warehouse we can distinguish between two modes of posing queries. *Preformulated* or *canned* queries have been formulated by the administrator and are run over and over again. They may have some variations by choosing parameters, but essentially they are fixed. Clearly, all the queries that are executed at the back end and in the core fall into this category, but canned queries also show up at the front end: graphical interfaces providing a managerial view of an organization and reports are based on canned queries.

Business analysts can also query a data warehouse in an *ad-hoc* mode. In an analysis session they formulate a series of related queries. For instance, in order to find out how a particular event influences the performance of the entire company, they will pose queries of increasing generality that touch more and more data. In a similar vein, to dig out the causes for a set of global statistical figures, they will proceed from general to specific queries.

Obviously, more time can be spent on the optimization of canned queries, since they are known long before their results are needed.

6.1.6 Multidimensional Queries

Among multidimensional queries, one can distinguish between those that retrieve data from the dimensions and others that retrieve factual information [Kimb96].

6.1.6.1 Querying the Dimensions

Before querying factual information from a data cube, an analyst will typically browse its dimensions, that is, use a query tool to find out the values of one or more attributes of a dimension, possibly while restricting other attributes. For example, the analyst may query the dimension *Product* for the values of the attribute *brand* when the attribute category is restricted to *dairy_products*.

If the data cube is implemented on a relational database according to the star model, then such a query logically consists just of projections and selections; if it is implemented as a snowflake, then the query also has to join the tables of the dimension in question.

Although dimension tables are by orders of magnitude smaller than fact tables, they can still be huge in some cases. The customer dimension in the data warehouse of a mail order company may well contain several million tuples. Therefore, browse queries, although structurally simple, require substantial optimization in order to be executed efficiently.

The goal of browsing is to identify interesting subsets of a dimension, called *constraint groups* because they satisfy certain constraints on the dimension, and to describe them by queries. In a query tool one can name such queries and store them. In a cooperative environment, constraint groups can also be made publicly available. In this case, the necessity arises to organize large sets of queries and to communicate their meaning. Tools support this by allowing one to attach comments to queries but do not offer any reasoning capabilities.

Other subsets of a dimension, called *behavioral groups*, are not solely definable in terms of the attributes of the dimension but involve also facts and restrictions on other dimensions. An example is the group of products that sell outstandingly well during the four weeks before Christmas. This group is defined not only in terms of the product dimension but also in terms of sales and time. Similar to constraint groups, there is also the necessity to manage queries defining behavioral groups.

6.1.6.2 Querying Factual Information

The ultimate goal of querying data cubes is to produce *business reports*. A business report is a table whose dimensions are labeled with values of attributes and whose facts are computed by aggregating facts in the underlying data cube. Often, the facts are not simple aggregates like counts and sums but are comparisons between different aggregates. In addition, a report may contain further aggregates like subtotals and totals, and exceptional values may be highlighted.

For computing a business report, there are two possibilities. The first is to translate the specification of the report into SQL. This poses problems when comparisons are to be computed. Comparing the sales of the current month with those of the previous month requires accessing the fact table at least twice. In SQL, this can either be achieved with a self-join or with correlated subqueries as in SQL-2. However, both options lead to code that is hard to write and for which query optimizers tend to produce inefficient execution plans. A further option is to distinguish between the selection criteria in a comparison (like "this month" and "last month" in our example) with a case statement as in SQL-92. Still, the resulting SQL code is cumbersome, and the selection constraints are torn apart because of the case analysis, which hampers optimization. The second option is to break down the report specification into a number of relatively simple SQL queries, one for each comparison, and to let the query tool assemble the results into the report. This strategy brings about several advantages. Since the queries are not too complex, basically select-project-join queries with aggregation, they can be executed efficiently by existing DBMSs. Also, incremental editing of the report is supported, since after a change, only those queries that are affected by it have to be recomputed.

Business reports can also be specified and computed with special OLAP servers. OLAP servers are specialized to store data cubes and to support multidimensional queries over them. They not only contain the basic cube, but also precomputed aggregates for the attributes of the dimensions. Thus, they guarantee fast response times for queries that do aggregation according to attributes. However, they often lack the flexibility to formulate more elaborate queries, or if they do, they may not run efficiently.

6.1.7 Extensions of SQL

Kimball addresses the shortcomings of SQL when decision support queries are to be formulated [Kimb96]. First, such queries require more sophisticated aggregation than is available in SQL (e.g., rank, *n*-tile, cumulative, moving-average, or moving-sum). Moreover, as seen before, for business queries comparisons are essential. If they are expressed in SQL or if a query tool produces the SQL code, they are executed by multiple sequential scans. This can be avoided if SQL is extended by a special syntax for comparisons, which can be evaluated more efficiently [ChRo96].

For specifying groups of aggregate queries, the operators *roll-up* and *cube* have been introduced as extensions to SQL in [GBLP97]. Each operator has as argument a relation and a list of attributes. It specifies a set of related queries that compute aggregates over groupings.

The *roll-up* operator aggregates over a series of groupings where each grouping is coarser than its predecessor. For example, roll-up for the list of attributes year, store, and price produces first an aggregation over the grouping by all the attributes, then by year and store, and finally by year alone. A roll-up query can be computed efficiently by deriving the result for one grouping from the result for its predecessor.

The operator *cube* groups by each subset of attributes from the list. Thus, if there are *n* attributes, cube specifies groupings. The groupings form a Boolean lattice that describes what can be computed from what. Contrary to roll-up, it is not straightforward to determine the most efficient strategy for computing the aggregates. Agrawal et al. report on an empirical analysis of different strategies for computing cube queries [AAD*96]. They extended existing grouping methods based on sorting and hashing with special optimization techniques such as combining common operations across multiple groupbys, caching, and reusing precomputed groupbys. The cube [GBLP97] does not presuppose hierarchies of attributes. However, a generalization to dimensions with hierarchies is straightforward: the cube operator groups by all sets of attributes where at most one attribute is taken from each hierarchy.

6.2 Query Processing Techniques

It is a characteristic of data warehouses that many queries are executed on the same set of data. This allows one to apply a broad spectrum of optimization techniques. In OLTP systems, auxiliary structures for answering queries like partitions of data, indexes, and materialized views must be continuously maintained. In the read-only environment of a data warehouse, they have to be updated only when the warehouse is loaded or refreshed (cf. Chap. 4). In the following, we will distinguish between three levels at which query execution can be optimized:

- Data access
- Evaluation strategies
- Exploitation of redundancy

6.2.1 Data Access

In any database, the time spent on reading data from secondary storage incurs the main cost of query processing. In data warehouses, this problem is aggravated, since the amount of stored data is enormous, and there is redundancy in the data. It is therefore important to have fast access to the data needed and to perform as much of the computation as possible on structures that consume only little space and therefore can be read and written quickly. These purposes are served by *indexes*. A detailed overview and performance comparison of index structures for data warehouses can be found in [Jürg02]. The following is a brief overview.

6.2.1.1 Indexes

An index on a relation *r* for an attribute *a* says where, for a value *v*, those records of *r* are stored that have *v* in position *a*. Traditional indexing technology relies on *B-trees*. In a B-tree, the location of a record is specified by a *row identifier* (RID). A B-tree for an attribute *a* of a relation *r* is a binary tree where each leaf node is labeled with a value *v* of *a* and with a list containing the RIDs of the records having *v* in position *a* (short RID *list*). Such indexes are also called *value list indexes*.

A value list index permits one to immediately access the records having a particular value for an attribute. However, if an attribute does not have many values, that is, if it is of low selectivity, and if the records with equal values are not clustered, it may be necessary to read the entire relation to get hold of the interesting records.

We call the set of records of *r* that satisfy a selection constraint the foundset of that constraint. Value list indexes can be useful for computing the RIDs of the foundsets of constraints that are given as Boolean combinations of value selections. If, for example, there are indexes for the attributes *a* and *b*, then the RIDs of the records satisfying *r.i* = *v* AND (OR) *r.b* = *w* can be obtained by taking the intersection (union) of the RID lists for the individual conditions.

For attributes with few distinct values (say: more than one hundred occurrences of the same values), *bitmap indexes* are more space efficient than value list indexes. Bitmap indexes differ from value list indexes in the way they represent the location of records: for each value *v* of *a*, there is a list of bits having 1 at position

i if the i-th record r_i satisfies $r_i.a = v$ and having 0 elsewhere. Because of the more compact representation, constraints on attributes can be more efficiently evaluated with bitmap indexes than with RID lists.

Bitmaps are well suited for data warehouses, since many attributes on dimensions have value sets of small cardinality. They can easily be generated at loading time. Browse queries and queries defining constraint groups on dimensions can be speeded up with them, since they essentially consist of Boolean combinations of value constraints for attributes. Moreover, they can be exploited for queries involving fact tables that constrain dimensional attributes.

Both value list and bitmap indexes can be useful to compute foundsets. However, to support operations on the values of some attribute a for the records in a foundset, projection indexes are more appropriate. A *projection index* of relation R for attribute a is a sequence whose i-th component is the value $r_i.a$ of the i-th record in r. In the presence of a bitmap for a foundset, a projection index can be used to retrieve the a-values of the records in the foundset.

Bitmap indexes are space-efficient for attributes with value sets of low cardinality but are inappropriate for attributes that take integers from a large range. In this case, *bit-sliced* indexes are applicable. Assume that the values of a are integers with $n + 1$ binary digits. Then we can conceptually decompose a into attributes $a_0,...,a_n$, each of which takes only 0 or 1 as values, so that $r.a = r.a_0 + 2 \cdot r.a_1 + ... + 2^n \cdot r.a_n$. For each $r.a_j$ one can create a bitmap index B^j. The bitmaps $B^0,..., B^n$ together form the *bit-sliced index* for a.

A bit-sliced index combines the properties of value-lists and simple bitmaps with those of projection indexes. It allows one both to locate the records with a particular value and to retrieve the value of a record at a given position.

Value-list, projection, and bit-sliced indexes can be used to calculate foundsets for range predicates of the form $n_1\ op_1\ r.a\ op_2\ n_2$, where n_1, n_2 are integers and op_1, op_2 are comparison operators (like <, =, etc.). An analysis in [ONQu97] shows that evaluation based on value list indexes performs best for narrow ranges, while the best results for wide ranges are obtained with bit-sliced indexes.

Data warehouses contain also large amounts of explanatory text, in the dimensions as well as in the metadata (e.g., text explaining the content of a table or the meaning of a rule), and so indexes supporting text search and information retrieval are important as well. Commercial products that support bitmap indexing are Model 204, TargetIndex by Redbrick, IQ by Sybase, and Oracle 7.3.

6.2.1.2 Aggregate Query Processing with Indexes

Indexes are also useful to answer aggregate queries. For illustration, we consider the simplest case: single column aggregates without grouping, like in

```
Select  sum(Sales.price)
From    Sales
Where   Condition
```

Suppose B^f is the bitmap of the foundset of *Condition*. Together with B^f, the indexes can be exploited in various ways to execute the query:

1. The records satisfying *Condition* are retrieved from disk by means of B^f; in each record, the price is looked up and added to the final sum.
2. The bitmap B^f is used to identify positions in a projection index for the attribute price. The values at those positions are added up.
3. Each leaf node in a value list B-tree is inspected; for each value, the number of records satisfying *Condition* is determined by intersecting the value list with B^f; from these numbers the sum can be calculated.
4. If $B^0,...,B^n$ is a bit-sliced index for price, then the sum looked for can be calculated by first computing the intersections $I^j := B^j \cap B^f$ of each slice with the foundset bitmap and then computing $\sum_i I_i^0 + I_i^1 \cdot 2^1 + ... + I_i^n \cdot 2^n$, where I_j^i is the i-th bit in the j-th bitmap.

O'Neil and Quass [ONQu97] show that the evaluation with bit-sliced indexes as in (4) is the method of choice for the aggregation functions *sum* and *avg*, while evaluation with value lists as in (3) is the most adequate for *max* and *min*. For *median* and *n-tile*, both index types allow for good performance. Techniques for computing aggregation queries can be generalized to aggregation with grouping and are meanwhile incorporated in several products.

6.2.1.3 Join-Indexes for Stars

When querying a star schema, indexes on a single table are of limited use. For instance, a condition like *day_of_week = Saturday* characterizes objects in a dimension – here those *Time* objects that fall on a Saturday – while we are ultimately interested in accessing entries in a fact table (e.g., the sales that happened on a Saturday).

The latter could be achieved by an index on the relation that results from joining the *Sales* fact table with the *Time* dimension table. A *join index* associates to the value v of an attribute a in a table R_1 (e.g., a dimension table) those records in a table R_2 (e.g., a fact table) that join with a record r_1 in R_1 that satisfies $r_1.a = v$.

Obviously, join indexes can be of any of the types discussed before: projection, value-list, bit-sliced. Join indexes can be applied in a similar way as the indexes on single relations discussed above. They allow one to evaluate queries on stars without actually performing joins and in some cases even without accessing the tables [ONGr95, ONQu97].

6.2.1.4 The Extended Data Cube Model

The data cube of a relation over several dimensions, as introduced in [GBLP96], is a primary structure for business analysis. For a relation r, it consists of the relation itself together with aggregates for all possible subsets of dimension attributes of r. Storing a data cube requires indexing, since the aggregates are usually very large. It turns out that using standard relational indexing requires enormous amounts of space and that the indexes are hard to maintain.

In order to implement data cubes efficiently, [RoKR97] have defined the extended data cube model (EDM) together with a special technique of mapping them onto R-trees. An extended data cube for a relation r with dimension attributes a, b, c and measure attribute m is a three-dimensional space, where each dimension consists of the possible values of the corresponding attribute, plus a special value, say "0." To some of the points in this space, a value from the domain of the measure is attached. In this structure, one can represent the original relation together with the aggregates: at the point with coordinates (a',b',c'), there is the value m' if r contains the tuple (a',b',c',m'); at the point with coordinates $(a',b',0)$, there is the value m'' if m'' is the aggregate for a', b' that is obtained when grouping by a and b, aggregating over c. In this model, many common OLAP-queries become range queries over the extended data cube, which makes the rich body of processing techniques for range queries applicable.

Kotidis and Roussopoulos [KoRo98] implement extended data cubes by mapping them to R-trees. In experiments they show that, compared to standard relational storage organization, their implementation uses less space. Most importantly, they succeed in significantly reducing the cost of query answering and updating the cube.

6.2.2 Evaluation Strategies

Rewriting a query into an equivalent form so that it is less expensive to evaluate and finding a plan for evaluating the query that incurs minimal cost are classical database problems. Data warehouses add some new aspects to these problems. Since queries involve huge sets of data, important speed-ups are possible if the data to be accessed can be reduced. Moreover, data warehouse queries require grouping and aggregation, for which additional techniques are needed. Finally, the schemata of relational data warehouses are usually laid out as stars or snowflakes, and thus have a restricted form, which allows one to apply specialized optimization methods. Numerous optimization issues require further study, e.g., cost-based vs. rule-based optimization, the timing of query optimization at compile time or run time, index selection and star join optimization, etc. This subsection focuses on two typical techniques for the OLAP setting:

- Interleaving group-by and join
- Optimization of nested subqueries

6.2.2.1 Interleaving Group-By and Join

Data warehouse queries over a star schema typically consist of joins of the fact table with the dimensions, of filter conditions on the dimensions, of grouping, and of aggregation. Traditional evaluation strategies for such queries schedule the join and filtering before the grouping and aggregation. However, early grouping and aggregation can reduce the size of intermediate results and thus reduce the cost of further processing. In addition, the scan for performing the join can be exploited for the grouping. Consider the following example [ChDa96]:

```
Select    Product.release_year, sum(Sales.price)
From      Sales, Store, Product
Where     Sales.pkey = Product.pkey AND
          Sales.skey = Store.skey AND
          Store.state = "California" AND
          Sales.year = 1996
GroupBy   Product.release_year
```

A traditional strategy for evaluating the query would be to first join the `Sales` fact table with its dimension tables `Store` and `Product` while filtering out the sales in California in 1996 and then to group by the `release_year`. A refined strategy is to aggregate by `Product.pkey` while computing the join of `Product` and `Sales`. The final grouping by `Product.release_year` can then be performed on a smaller intermediate relation.

Work on interleaving group by and join has come up with transformation rules that move and add group-by operators in queries [Daya87, GuHQ95, ChSh94, YaLa94, YaLa95]. Chaudhuri and Shim [ChSh94] have shown how to integrate these transformations into a traditional query optimizer.

6.2.2.2 Optimization of Nested Subqueries

Queries produced by end user interfaces usually involve comparisons, even comparisons with aggregates. In SQL, such queries are expressed by nesting queries and subqueries. Nested queries can sometimes be rewritten into single block queries, which can be executed more efficiently (see, e.g., [Daya87]).

Also in cases where the queries cannot be rewritten into a single block, optimization is possible by moving constraints around and by suitably ordering the subqueries. We illustrate this optimization technique by an example, taken from [ChDa96]:

> *Find all employees younger than 35 who earn more than the average of their department.*

The most straightforward evaluation plan is for each employee to (1) check whether the employee is younger than 35, (2) find the department of the employee, (3) compute the average salary in the department, and (4) check whether the employee's salary is above the average. This is obviously inefficient, since the average salary for a department may be computed many times. Therefore, the following is a better evaluation plan: (1) for each department, compute the average salary, and (2) for each employee younger than 35, check whether the salary is above the average. This plan may involve too much processing if there are many departments that have only senior employees. It is therefore more efficient to perform Step (1) only for those departments that have an employee of age under 35.

6.2.3 Exploitation of Redundancy

Data warehouse queries, in particular those issued by end users, involve aggregation, which is expensive to compute. The main technique for speeding up such queries is to precompute aggregate views and to materialize them, as Kimball [Kimb96] said about the design of commercial data warehouses: "The use of prestored summaries (aggregates) is the single most effective tool the data warehouse designer has to control performance."

In order to apply this technique successfully, one has to answer the following questions:

- Which views are useful for answering a query?
- What is the expected size of an aggregate view?
- Which aggregate views should be precomputed?

Research on the first two questions is discussed in the present chapter. For a discussion of the third question we refer to Sects. 7.5 and 8.3.

6.2.3.1 Which Views are Useful for Answering a Query?

This question, known as the view usability problem, has received much attention in the past few years. In an abstract manner, it can be stated as follows:

> *Given views $V_1,...,V_n$ and a query Q over a fixed database schema, can Q be reformulated using the views so that it does not use any (or as few as possible) of the relations in the database?*

The problem is parameterized by the languages in which views and queries are expressed and by the semantics under which they are evaluated. Special cases of the view usability problem are the *equivalence problem* (do two queries Q_1, Q_2, always produce the same set of answers?) and the *containment problem* (is the set of answers to Q_1 always a subset of those to Q_2?). For query languages as expressive as relational calculus the view usability problem is obviously undecidable, since the equivalence of relational calculus queries is undecidable. Therefore, attention has been focused on more restricted cases.

Containment and equivalence under set semantics have been studied extensively for *conjunctive queries*, also known as select-project-join (SPJ) queries [ChMe77, AhSU79, JoKl83, SaSa92], for conjunctive queries with built-ins [VaDM92, LeSa95], for queries with union and difference [SaYa80], and for conjunctive queries defined by Datalog programs [LeSa95, LMSS93]. Containment for conjunctive queries under multiset semantics as in SQL has been investigated in [ChVa93].

Reasoning about containment of queries is not only relevant to determine views that can be used for answering a query. It can be applied to organize large sets of queries into taxonomies. In a data warehouse environment, this can be important to support navigation among constraint groups, which are defined by SPJ queries, or behavioral groups. It can also help a user to find constraint groups similar or related to ones he is interested in, which can be particularly difficult if the groups have been defined by different users.

Techniques for using views to answer queries have been suggested by a number of researchers [YaLa87, ChRo94, CKPS95], although most did not pay much attention to the formal aspects of the problem. A related question is how to use cached results of previous queries to answer new queries and to determine which results to cache [Sell88, SJGP90]. They determine usability syntactically, by common expression analysis [Fink82]. In the successful commercial system ADMS [ChRo94], not only final results, but also intermediate results corresponding to inner nodes in the query tree are cached.

The view usability problem for conjunctive queries under set semantics has been treated in [LMSS95]. Like equivalence and containment, view usability is NP-complete. The algorithm in [LMSS95] generates for a query and a set of views all possible SPJ queries over the views that are equivalent to the original query. The algorithm can also take into account order constraints but no grouping and aggregation. Chaudhuri et al. investigated view usability for conjunctive queries under multiset semantics, in [CKPS95], and they integrated a view usability algorithm into a cost-based optimizer.

A method to use views for queries with grouping and aggregates has been developed by [GuHQ95]. The method is based on rewriting rules to transform the tree representation of a query. The method is sound, but it does not allow one to find all possible equivalent rewritings using the views.

Dar et al. studied the same problem and gave sufficient conditions for an aggregate SQL-query to be computable from a set of views. Their algorithms are guaranteed to be complete in some cases, e.g., when the views do not contain aggregation and all the constraints in the where part of the query and of the views contain only equality predicates [DJLS96].

Chaudhuri and Dayal have given sufficient conditions for aggregate views to be usable to answer queries over a star schema [ChDa96]. He distinguishes between exact and nonexact matches. The simplest views and queries have form of the following view V:

```
Select     Product.brand_name,
           Product.brand_introduced,
           sum(Sales.price)
From       Sales, Product
Where      Sales.pkey = Product.pkey AND
           Product.brand_introduced > 1990
GroupBy    Product.brand_name.
```

They consist of (1) joins of a fact table with its dimension tables (here `Sales` and `Product`), (2) a grouping by dimension attributes (here `brand_name`), and (3) aggregation of numeric measures in the fact table (here `price`). It is easy to see that the query Q_1, defined by the SQL statement below, can be computed using V.

```
Select     Product.brand_name,
           Product.brand_introduced, sum(Sales.price)
From       Sales, Product
Where      Sales.pkey = Product.pkey AND
           Product.brand_introduced > 1990
GroupBy    Product.brand_name.
```

The answers to Q_1 are obtained by choosing only those answer records of V, where `brand_introduced > 1991`, and then projecting out the column `brand_introduced`. More abstractly, there is an *exact match* from Q_1 to V, because (1) each projection column of the query is present in the view, (2) the aggregation functions on each measure match, (3) each selection constraint in the query implies a constraint in the view, and (4) the attributes occurring in query constraints that are strictly stronger than constraints in the view are also present in the view, so that they can be used to sharpen those constraints. For the query Q_2 below there is no exact match, since for instance `product_name` is not among the projection columns in V:

```
Select    Product.product_name, sum(Sales.price)
From      Sales, Product
Where     Sales.pkey = Product.pkey AND
          Product.brand_introduced > 1991
GroupBy   Product.product_name.
```

Nonexact matches take also the granularity of grouping into account. Consider, for instance, a view V that contains the

> *Sum of sales for each product and for each year, where the product has been released after 1990,*

and a query Q_1 asking for the

> *Sum of sales for each brand and for every 5 years, where the product has been released after 1992.*

There is no exact match, but Q_1 can be computed from V by further aggregation, since the grouping is coarser and the constraints are more restrictive. However, there is no such match for the following query Q_2.

> *Sum of sales for each product and for each month, where the product has been released after 1992,*

since in Q_2 the grouping on time is finer than in V.

The first work on aggregate queries that gives not only sufficient conditions for problems related to view usability is [NuSS98], who provided characterizations for the equivalence of aggregate queries. They investigated conjunctive queries with comparisons and the aggregate operators *min, max, count, count-distinct,* and *sum.* Essentially, this class contains all unnested SQL queries with the above aggregate operators, with a where-clause consisting of a conjunction of comparisons and without a `Having`-clause. The characterizations differ, depending on the aggregate operators, on the absence or presence of comparisons, and on domain over which the comparisons are interpreted. All characterizations are decidable with polynomial space. For the special case of linear queries, i.e., queries with no repeated predicates in their bodies, equivalence can be decided in polynomial time.

View usability for queries with aggregation is one of the core problems in query processing over data warehouses. There is a continuous stream of contributions to this topic. So far, however, the various contributions sum up only to a very fragmented picture. There is no general statement of the problem, which is partly due to the fact there is no adequate formalization of a multidimensional data model and query language. As a consequence, it is unclear whether or not the greatest part of the problem has already been solved. However, further problems are still open. One of them is how to use more than one view to answer a query, e.g., to compute summaries from averages and cardinalities. Another one, which is currently getting more attention, is view reuse in semistructured data models [CDLV00a-c].

A very important step to carry these results into practice is the integration of view reuse strategies with commercial, quantitatively oriented query optimizers. Microsoft has recently developed a solution for their SQL Server which supports rapid retrieval of a small set of reuse candidates among potentially thousands of materialized views through clever indexing of view definitions. It then uses the normal query optimizer to make the choice between these candidates [GoLa01]. The class of queries covered by this approach is so-called SPJG queries on relational databases, i.e. select-project-join queries with a single group-by at the end. Performance studies show very short optimization times and impressive query speed-up even in the presence of thousands of materialized views where, in other approaches, optimization effort would be prohibitive.

6.2.3.2 What is the Expected Size of an Aggregate View?

The precomputation of aggregates may lead to a disproportional storage blow up. In order to estimate the benefit of materializing an aggregate view, it is therefore necessary to have a good estimate of its size without computing it.

If the grouping attributes of the aggregation are statistically independent, then the size of the view can be estimated using simple combinatorics: it depends on the number of distinct values of the attributes. Sometimes, these numbers are available through system statistics. If not, they can be estimated by sampling techniques [HNSS95].

The combinatorial approach overestimates the view size if the values of the grouping attributes display some dependency, i.e., intuitively speaking, if the tuples are clustered in the data cube. A crude estimate can be obtained in this case, by computing the view only for a fraction of the data and then linearly scaling the result to the size of the entire database. More sophisticated methods are based on probabilistic counting. Experiments show that the latter yields the most precise estimates at a cost that is linear in the size of the database [SDNR96].

6.3 Conclusions and Research Directions

Query optimization research has concentrated on queries in the relational model. For data warehouses, the semantics offered by the relational model seem to be insufficient. Data warehouse queries – in particular those issued at the back end – are usually very complex. For instance, with a query tool, an end user defines a report that is broken down into several relational queries. If only the relational queries are considered as the target of optimization, the dependencies between the components of a report cannot be taken into account.

For this reason, queries in data warehouses cannot be adequately dealt within a purely relational framework. Conceptually, data warehouses implement some multidimensional data model. This model can be implemented on relational databases (ROLAP), and it can also be implemented on some dedicated multidimensional architecture (MOLAP). Many of the problems in query optimization described earlier can be treated at the more abstract level of the multidimensional data model. This will give an advantage, since results will hold for ROLAP and MOLAP as well.

Even on a relational DBMS, one multidimensional database can be implemented in various ways, e.g., as a proper star or as a snowflake. In a relational setting, one assumes that queries are relations derived from other relations. For a data warehouse this need not be true. There are conceptual views, e.g., aggregates computed from some basic fact data, which are stored in the same table as the fact data. They can only be recognized by looking at the values of particular attributes that indicate the "level" of aggregation. This is one of the motivations for the efforts, described in Chap. 5, to thoroughly formalize multidimensional data models. Many further results in this important area continue to emerge.

7 Metadata and Data Warehouse Quality

In the traditional view, data warehouses provide large-scale caches of historic data. They sit between (a) information sources gained externally or through online transaction processing systems (OLTP) and (b) decision support or data mining queries following the vision of online analytic processing (OLAP). Three main arguments have been put forward in favor of this caching approach:

1. *Performance and safety considerations.* The concurrency control methods of most DBMS do not react well to a mix of short update transactions (as in OLTP) and OLAP queries that typically search a large portion of the database. Moreover, the OLTP systems are often critical for the operation of the organization and must not be in danger of corruption by other applications.
2. *Logical interpretability problems.* Inspired by the success of spreadsheet techniques, OLAP users tend to think in terms of highly structured multidimensional data models, whereas information sources offer at best relational, often just semistructured data models or even flat files.
3. *Temporal and granularity mismatch.* OLTP systems focus on current operational support in great detail, whereas OLAP often considers historical developments in a somewhat lesser detail.

Thus, quality considerations have accompanied data warehouse research from the beginning. As shown in the previous chapters of this book, a large body of practical experience and research literature has evolved over the past few years in addressing the problems introduced by the DW approach, such as the trade-off between freshness of DW data and disturbance of OLTP work during data extraction; the minimization of data transfer through incremental view maintenance; and a theory of computation with multidimensional data models.

However, the heavy use of highly qualified consultants in data warehouse applications indicates that we are far from a systematic understanding and usage of the interplay between quality factors and design options in data warehousing. The goal of the European DWQ project [JaVa97] was the engineering of solutions and methods towards these issues by developing, prototyping, and evaluating comprehensive *Foundations for Data Warehouse Quality*, delivered through *enriched metadata management facilities* in which specific analysis and optimization techniques are embedded.

After giving a short overview of the state of the practice in handling data warehouse quality, this chapter further develops the DWQ architecture and quality management framework introduced in Sect. 2.7 and links it to other work on data and software quality. The chapter ends with a detailed example where the approach is used to analyze the quality of data staging techniques.

7.1 Metadata Management in Data Warehouse Practice

Metadata play an important role in data warehousing. Before a data warehouse can be accessed efficiently, it is necessary to understand what data is available in the warehouse and where is the data located. In addition to locating the data that the end users require, the metadata may contain the following [AdCo97, MStr95, Micr96]:

- Data dictionary: Contains definitions of the databases being maintained and the relationships between data elements
- Data flow: Direction and frequency of data feed
- Data transformation: Transformations required when data is moved
- Version control: Changes to metadata are stored
- Data usage statistics: A profile of data in the warehouse
- Alias information: Alias names for a field
- Security: Who is allowed to access the data

Metadata is stored in a repository, where it can be accessed from every component of the data warehouse. Because metadata is used and provided by all components of the warehouse, a standard interchange format for metadata is necessary. The Metadata Coalition (MDC) has proposed a *Metadata Interchange Specification* [MeCo96]; additional emphasis has been placed on this area through the recent efforts of Microsoft to introduce a repository product in their Office suite, including some simple information models for data warehousing [BBC*99]. In addition, a number of metadatabase systems developed in research have been successfully used in industry. In the following subsections, four such approaches are described.

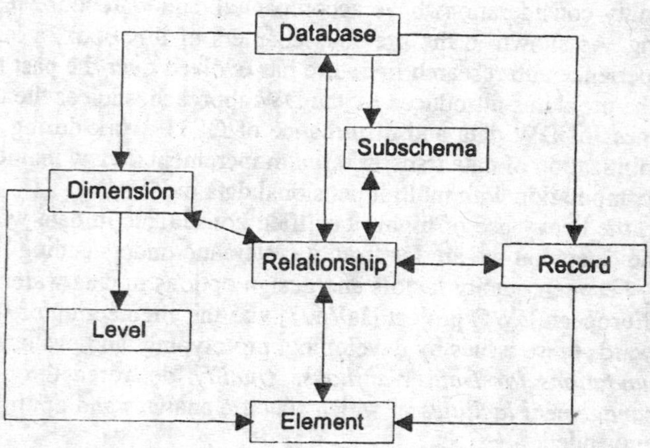

Fig. 7.1. MDIS metamodel [MeCo96]

7.1.1 Metadata Interchange Specification (MDIS)

The MDC is an open group of companies including IBM, Sybase, Informix, and Prism Solutions. The goals of the MDC are to create standard access mechanisms and a standard application programming interface for metadata, to enable users to control and manage the access and manipulation of metadata in their environments through the use of interchange specification-compliant tools, and to define a simple interchange implementation infrastructure that will facilitate compliance by minimizing the amount of modification to existing tools.

The metadata interchange specification (MDIS) is character-based so that it is platform independent. One of the most important features is the extensibility of the specification, because the contents of what is considered metadata will be evolved. The metamodel of MDIS describes entities and relationships that are used to represent metadata and is illustrated in Fig. 7.1.

A *database* object represents a database system or a group files. A database consists of several *records* (e.g., tables in a relational database). The purpose of a record is to provide a physical grouping of element objects that describe a unit of data. An *element* is the smallest piece of data that can be represented. For example, the element dept_name of the record department represents the name of a department.

A *subschema* describes a logical grouping of record objects. A *Relationship* object represents a relationship between object types. A relationship has a type, which may be EQUIVALENT, DERIVED, INHERITS-FROM, CONTAINS, INCLUDES, LINK-TO, or USER-DEFINED. Relationships are used to express join relationships between tables (i.e., referential integrity) in relational databases or inheritance between two record objects in object-oriented databases.

Dimensions are used to represent dimension tables in a multidimensional database. The *level* object gives the position of the dimension in the dimension hierarchy. For example, the dimension *Product* is in level 1 and the dimension *Product-Line* is in level 2 and so on.

Each specification of metadata also has a *header* in which some general information about the data is stored (e.g., creation data, exporting tool, version of tool and MDIS, character set). The definitions are exported as ASCII texts so that they can be read on all platforms. The example in Fig. 7.1 describes a relational database.

7.1.2 The Telos Language

Most of the current repository or metadata tools rely on relational architectures. This is often considered insufficient to handle complexities of the different views of metadata integrated with disparate sources. Objectification of this architecture may offer a much more flexible environment both for data warehousing and other needs for a true corporate metadata model [Sach97].

The Telos language developed jointly between the University of Toronto and a number of European projects in the late 1980s is specifically dedicated to this goal [MBJK90]. Telos, in the axiomatized form defined in [Jeus92], offers an unlimited classification hierarchy in combination with abstractions for complex objects and for generalizations, both for objects and for links between them, i.e., both are first-

```
BEGIN HEADER
      CharacterSet 'ENGLISH'
      ExportingTool 'DB2'
      ToolVersion '6.5'
      ....
END HEADER

BEGIN DEFINITION
COMMENT Representing a relational database in MDIS
END DEFINITION

BEGIN DATABASE
      Identifier 'EINSTEIN.SYSADMIN.COURSE_CATALOG'
      DateCreated '1995-04-12'
      BriefDescription 'DB2 database containing department ...'
      ServerName 'EINSTEIN'
      OwnerName 'SYSADMIN'
      DatabaseName 'COURSE_CATALOG'
      DatabaseType 'RELATIONAL'
      ...
      COMMENT MDIS description of tables
      BEGIN RECORD
            Identifier 'EINSTEIN.SYSADMIN.COURSE_CATALOG.DEPT'
            BriefDescription 'Record describing department'
            RecordName 'DEPT'
            ...
            COMMENT MDIS description of an attribute
            BEGIN ELEMENT
                  Identifier '...COURSE_CATALOG.DEPT.DEPT_NAME'
                  BriefDescription 'Name of department'
                  ElementDataType 'VARCHAR'
                  ...
            END ELEMENT
            COMMENT more elements and records ...
      END RECORD
END DATABASE

COMMENT Relationship between dept and course
BEGIN RELATIONSHIP
      Identifier '...DEPT.DEPT_ID<EQUIVALENT>...COURSE.DEPT_ID'
      RelationshipName 'Dept-Course'
      SourceObjectIdentifier '...DEPT.DEPT_ID'
      TargetObjectIdentifier '...COURSE.DEPT_ID'
      RelationshipType 'EQUIVALENT'
      RelationshipOrdinality '1:N'
      ...
END RELATIONSHIP
```

Fig. 7.2. MDIS example definition of a relational database [MeCo96]

class citizens of the language, offering maximum flexibility in modeling and remodeling complex metadata. In particular, it becomes possible to define, not just syntactically but also semantically, metamodels for new kinds of metadata introduced in the distributed system managed by the repository, and therefore in the repository itself. Such metamodels are often also called *Information Models* and exist typically for all kinds of objects and processes used in system analysis, design, and evolution.

Two Telos implementations have found rather widespread use in research and industry.

The ConceptBase system, developed since 1987 at RWTH Aachen, Germany [JaRo88, JGJ*95, NiJa98], as a knowledge-based repository, integrates a logical semantics with the basic abstractions for the purpose of analyzing the consistency of stored repository objects such as software specifications in different kinds of formalisms. Through the axiomatization of the Telos semantics, recently extended to the case of distributed metadata [NiJa98], ConceptBase achieves a combination of structural object-orientation with the kinds of reasoning capabilities offered by deductive relational databases; this combination is exploited in the implementation which combines a special-purpose object store with relational query optimization techniques, including deduction, which has proven rather important in quality analysis, as will be shown later in this chapter.

The Semantic Index System developed in the early 1990s at the FORTH Institute in Heraklion, Greece [CJMV95], has been used mainly for semantic indexing of large collections multimedia objects. In contrast to the deduction-oriented approach of ConceptBase, the focus lies on making the basic structural abstraction mechanisms of Telos available in a highly reliable and scalable system, implemented directly on top of a special-purpose object store.

7.1.3 Microsoft Repository

The *Microsoft Repository Version 2 (MSR)* [BBC*99], currently marketed under the name of Meta Data Engine, can also be considered a combination of relational and object-oriented solutions. While its underlying storage mechanism is relational, the data model is based on Microsoft's Common Object Model (COM), a binary object standard.

As in the case of Telos, a main strategy of MSR has been the definition of a broad range of information metamodels for system domains relevant for Microsoft customers and object exchange models (OEMs). However, while Telos employs logical formalisms to define the relationships between repository objects, MSR needs to do this with structural information and some object-oriented methods and abstraction mechanisms. Microsoft has therefore decided to standardize all metadata schemas (information models) within the context of a predefined metamodel of Rational's Unified Modeling Language (UML).

Within this framework, MSR also supports a number of repository Information Models and services directly targeted at Data Warehousing, such as the data cube offered in the OLAP extensions of Microsoft SQL Server [GBLP97] and some typical source-to-warehouse transformations. In addition, the application-independent kernel of MSR offers additional features such as fine-grained version control.

7.1.4 OIM and CWM

Building on experiences with the approaches described above, two industry standards have emerged in recent years: the Open Information Model (OIM) by the already mentioned MDC and the Common Warehouse Metamodel (CWM) by the Object Management Group (OMG). A detailed comparison of both approaches can be found in [VeVS99].

In essence, the goal of OIM [MeDC99] was to support lifecycle-wide tool interoperability. The portion of OIM focused on data warehousing addresses the description of static aspects such database schema elements (following the SQL standard), OLAP schema elements, record-oriented database elements, and report definitions, as well as a hierarchy of dynamic aspects including data transformation maps, transformation steps, and transformation packages. The MDC OIM uses UML both as a modeling language and as the basis for its core model. The OIM is divided in submodels, or *packages*, which extend UML in order to address different areas of information management. The *Database and Warehousing Model* comprises the Database Schema Elements package, the Data Transformations Elements package, the OLAP Schema Elements package, and the Record Oriented Legacy Databases package. The *Database Schema Elements* package contains three other packages: a *Schema Elements* package (covering the classes modeling tables, views, queries, indexes, etc.), a *Catalog and Connections* package (covering physical properties of a database and the administration of database connections), and a *Data Types* package, standardizing a core set of database data types.

The *Data Transformations Elements* package (cf. Fig. 7.3) covers basic transformations for relational-to-relational translations. The package does not deal with data warehouse process modeling (i.e., it does not cover data propagation, cleaning rules, or the querying process) but covers in detail the sequence of steps, the functions and mappings employed, and the execution traces of data transformations in a data warehouse environment. A *transformation* maps a set of sources to a set of target objects, both represented by a *transformable object set* (typically sources and targets are columns or whole tables). A transformation has a *Function Expression* property to provide a description of the executed code/script. A *transformation task* describes a set of transformations that must be executed together-a logical unit of work. A *transformation step* executes a single task and is used to coordinate the flow of control between tasks. The distinction between task and step is that a step is a logical unit of executions (i.e., it could also belong to the physical perspective) and a task is a logical unit of work (i.e., it is a logical list of transformations, which fulfill the atomicity property; a task is also characterized by a rollback *inverse transformation*). A *Step Precedence* is a logical connector for steps: a step can be characterized by its *preceding* and *succeeding* step precedence instance. The step precedence indicates that the successor steps can only be executed when all preceding steps have committed – i.e., it is a join-and node in the Workflow Management Coalition (WfMC) terminology. A *transformation package* is the unit of storage for transformations and consists of a set of steps, packages, and other objects. *Package executions* express the concept of data lineage, covering the instances of the *step executions*.

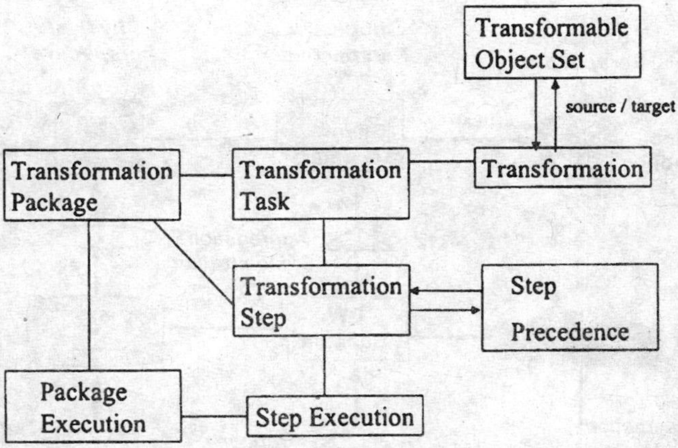

Fig. 7.3. The Microsoft proposal for Data Transformation Elements [MeDC99]

The CWM is narrower but provides more support for data warehousing itself. In addition, it is distinguished from the OIM by relying on object-oriented and semistructured modeling technologies such as UML, XML, and CORBA. It is organized in a foundational metamodel comprising business information, data types and CWM types, expressions, keys, and indexes, on which a number of more specific model packages for both MOLAP and ROLAP solutions are based. These address warehouse deployment in terms of hardware and software, packages for accessing relational, XML-based, and record-oriented sources, packages for multidimensional and relational data warehouses to reach OLAP functionality, and process-oriented packages comprising individual transformations, process flow, and day-to-day operation.

7.2 A Repository Model for the DWQ Framework

In Sect. 2.7, we argued the need for a data warehouse metadata structure that offers three perspectives: a conceptual business perspective with the enterprise model at the center, a logical perspective with the data warehouse schema at the center, and a physical perspective representing the physical data transport (e.g., in the query processing and data refreshment process). Each of these perspectives and their interrelationships is orthogonally linked to the three traditional layers of data warehousing, namely sources, data warehouse, and clients. The framework is reproduced in Fig. 7.4.

In this section, we elaborate the extended metamodel resulting from this approach and show how it can be implemented in a repository. The application of these repository concepts is illustrated with a more detailed description of a few specific submodels developed.

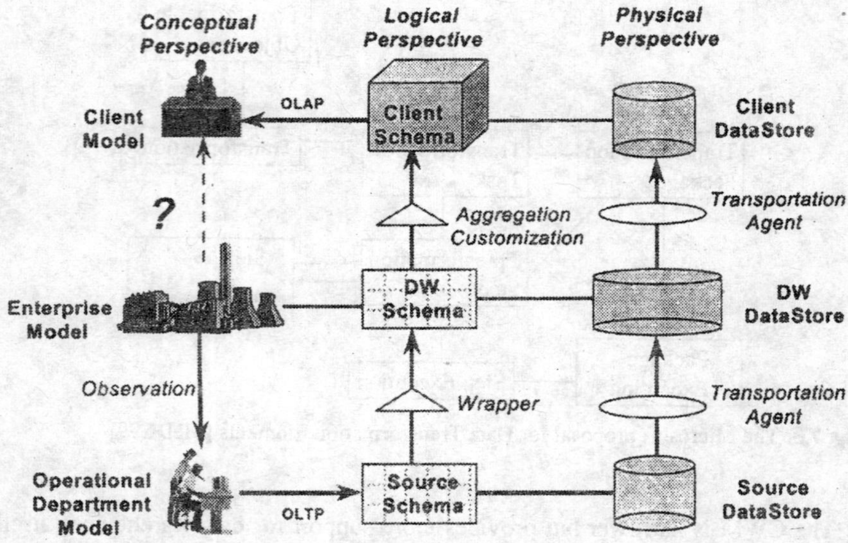

Fig. 7.4. The DWQ data warehouse metadata framework

We use the metadatabase to store an abstract representation of data warehouse applications in terms of the three-perspective scheme. The architecture and quality models are represented in Telos [MBJK90] (see Sect. 7.1.2), an extensible metamodeling language that has a graphical syntax and a frame syntax, both mapped to an underlying formal semantics based on standard deductive databases. Using this formal semantics, the Telos implementation in the ConceptBase system [JGJ*95] provides query facilities and definition of constraints and deductive rules. Telos is well suited because it allows the formalization of specialized modeling notations (including the adaptation of graphical representations) by means of metaclasses. Since ConceptBase treats all concepts including metaclasses as first-class objects, it is well suited to manage abstract representations of the DW objects to be measured.

A condensed ConceptBase model of the architecture notation is given in Fig. 7.5, using the graph syntax of Telos. Bold arrows denote specialization links. The top-level object is *MeasurableObject.* It classifies objects at any perspective (conceptual, logical, or physical) and at any level (source, data warehouse, or client). Within each perspective, we distinguish between the modules it offers (e.g., client model) and the kind of information found within these modules (e.g., concepts and their subsumption relationships). The horizontal links *hasSchema* and *isViewOn* exemplify how the horizontal links in Fig. 7.4 are interpreted: the types of a schema (i.e., relational or multidimensional structures) are defined as logical views on the concepts in the conceptual perspectives. On the other hand, the logical structure of the components of the physical perspective is described from the respective *Types* in the logical perspective, through the relationship *hasStructure*.

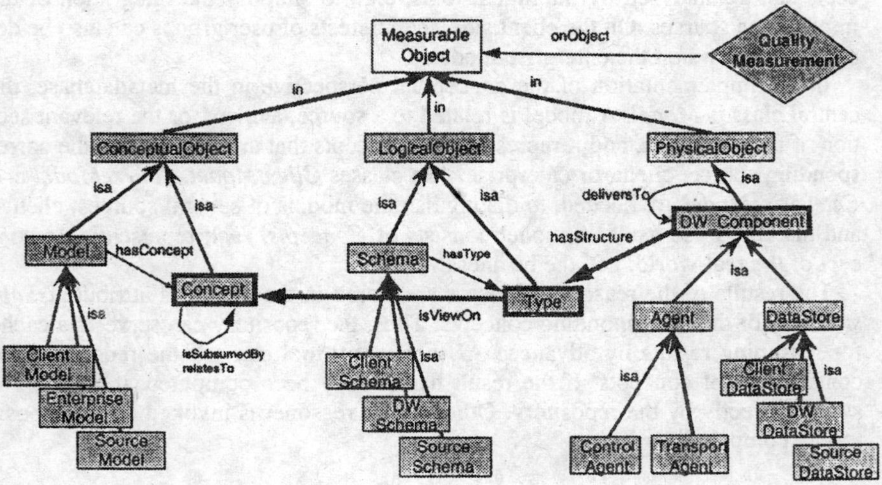

Fig. 7.5. Structure of the repository metamodel

Each object can have an associated set of (materialized) views called *Quality-Measurements.* These materialized views (which can also be specialized to the different perspectives – not in Fig. 7.5) constitute the bridge to the quality model discussed later.

The horizontal levels of the objects (i.e., *Source, Data Warehouse,* and *Client*) are coded by the three subclasses attached to *Model*, *Schema*, and *DataStore*. We have experimented with this notation and were able to represent physical data warehouse architectures of commercial applications, such as the SourcePoint tool marketed by Software AG, the DW architecture underlying a data mining project at Swiss Life [KiRS97, StKR98] and a DW project in Telecom Italia [TrLN99]. The logical perspective currently supports relational schema definitions whereas the conceptual perspective supports the family of extended entity-relationship and similar semantic data modeling languages. Note that all objects in Fig. 7.4 are metaclasses: actual conceptual models, logical schemas, and data warehouse components are represented as instances of them in the metadatabase.

In the following subsections, we elaborate on the purpose of representing each of the three perspectives, and then demonstrate how the architecture above can be refined for particular purposes.

7.2.1 Conceptual Perspective

The conceptual perspective is a view on the business model of the information systems of an enterprise. The central role is played by the enterprise model, which gives an integrative overview of the conceptual objects of an enterprise. The models of the client and source information systems are considered as views on the enterprise model, i.e., their contents are described in terms of the enterprise model. One goal of the conceptual perspective is to have a model of the information independent from physical organization of the data, so that relationships between concepts can be analyzed by intelligent tools, e.g., to simplify the integration of the information sources. On the client side, the interests of user groups can also be described as views on the enterprise model.

In the implementation of the conceptual perspective in the metadatabase, the central class is *Model*. A model is related to a source, a client, or the relevant section of the enterprise, and it represents the concepts that are available in the corresponding source, client, or enterprise. The classes *ClientModel*, *SourceModel* and *EnterpriseModel* are needed, to distinguish the models of several sources, clients, and the enterprise itself. A model consists of *Concepts*, each representing a concept of the real-world, i.e., the business world.

The results of the reasoning process are stored in the model as attribute *isSubsumedBy* of the corresponding concepts. Thus, the repository can serve as a cache for reasoning results by advanced AI tools. Any tool can ask the repository for containment of concepts. If the result has already been computed, it can be answered directly by the repository. Otherwise, a reasoner is invoked by the repository to compute the result.

7.2.2 Logical Perspective

The logical perspective conceives a data warehouse from the viewpoint of the actual data models involved. The central concept in the logical perspective is *Schema*. As a model consists of concepts, a schema consists of *Types*. We have implemented the relational model as an example for a logical data model; other data models such as the multidimensional or the object-oriented data model are also being integrated in this framework [GeJJ97, Vass98].

Analogical to the conceptual perspective, we distinguish among *ClientSchema*, *DWSchema*, and *SourceSchema* for the schemata of clients, the data warehouse, and the sources. For each client or source model, there is a separate corresponding schema. This restriction is guaranteed by a constraint in the architecture model. The link to the conceptual model is implemented through the relationship between concepts and types: each type is expressed as a view on concepts.

7.2.3 Physical Perspective

The data warehouse industry has mostly explored the physical perspective, so that many aspects in the physical perspective are taken from the analysis of commercial data warehouse solutions such as Software AG's SourcePoint tool, the data warehouse system of RedBrick, Essbase of Arbor Software, or the product suite of MicroStrategy (see Chap. 1). We have observed that the basic physical components in a data warehouse architecture are *agents* and *data stores*. *Agents* are programs that control other components or transport data from one physical location to another. *Data stores* are databases which store the data that is delivered by other components.

The basic class in the physical perspective is *DW_Component*. A data warehouse component may be made up of other components. This fact is expressed by the attribute *hasPart*. Furthermore, a component *deliversTo* another component a *Type*, which is part of the logical perspective. Another link to the logical model is the attribute *hasSchema* of *DW_Component*. Note that a component may have a schema, i.e., a set of several types, but it can only deliver a type to another component. This is due to the observation that agents usually transport only "one tuple at a time" of a source relation rather than a complex object.

One type of component in a data warehousing environment is an *Agent*. There are two types of agents: *ControlAgent*, which controls other components and agents (e.g., it *notifies* another agent to start the update process) and *TransportationAgent*, which transports data from one component to another component. An *Agent* may also notify other agents about errors or termination of its process.

Another type of component is a *DataStore*. It physically stores the data which is described by models and schemata in the conceptual and logical perspective. As in the other perspectives, we distinguish among *ClientDataStore*, *DW_DataStore* and *SourceDataStore* for data stores of clients, the data warehouse, and the sources.

7.2.4 Applying the Architecture Model

The metadata framework shown in Fig. 7.5 defines the basic metamodel of the products in the repository and their interrelationships. As shown in Fig. 7.6, this framework can be instantiated by information models (conceptual, logical, and physical schemas) of particular data warehousing strategies which can then be used to design and administer the instances of these data warehouses.

However, quality cannot just be assessed on the network of nine perspectives but is largely determined by the *processes* how these are constructed. The process metamodel defines how such processes can be defined. The process models define plans on how data warehouse construction and administration is to be done, and the traces of such processes are captured at the lowest level. This process hierarchy accompanying the DW product model is shown on the right of Fig. 7.5. An example of such a process was given in Chap. 4 with the data refreshment process.

Based on earlier work on quality-oriented software process modeling and management [JaPo92] and under consideration of the workflow modeling standards defined by the Workflow Management Coalition, a conceptual model for data warehouse process management has been defined in [VQVJ01]. This model describes how human or computerized agents use and produce information at the conceptual, logical, and physical levels, thus customizing the basic flow of information from sources to clients. Even though this process is conceptually simple, the interplay of conceptual, logical, and physical aspects can get extremely intricate in practice.

This challenge to metadata management is probably best illustrated by Fig. 7.7 which describes the basic information flow in the data warehouse for financial controlling in a large international bank [ScBJ00]. Figure 7.7 lists no less than sixteen submodels addressing all three perspectives, reflecting the difficulties of reconciling different kinds of financial semantics, heterogeneity of source data models and source availability, bottlenecks involved in scheduling huge amounts of data flows during daily refreshment, and personalization of client interests beyond their role definitions.

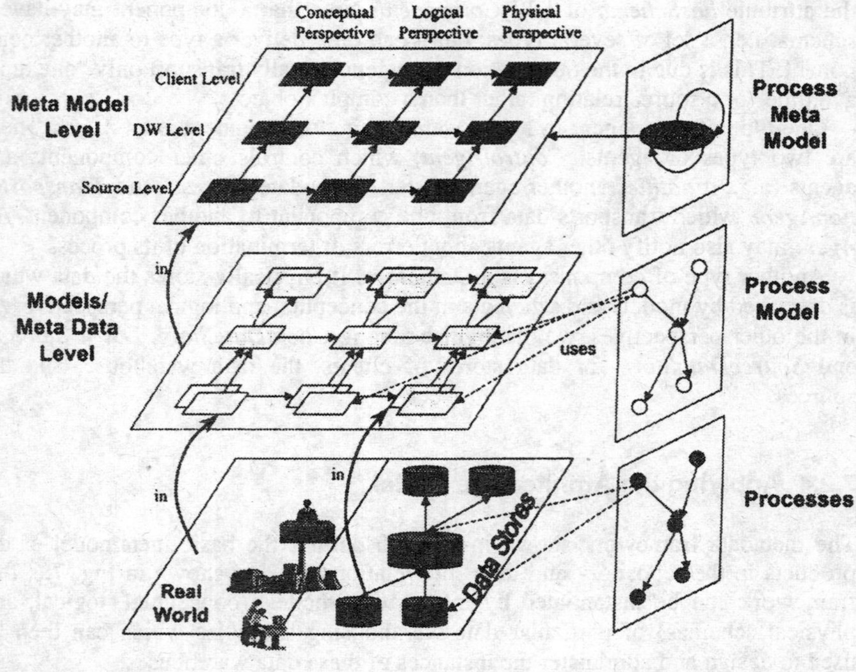

Fig. 7.6. Repository structure for capturing product and process of data warehousing

For the purposes of this chapter, we assume that the impact of such process models on the repository is some kind of query plan, i.e., a partially ordered set of queries defined over the metadatabase (and stored in the metadatabase). This is, for example, also the strategy followed in the Microsoft Repository [BBC*99].

In the remainder of this subsection, we give an example of how our approach can be applied to describe a specific task and solution strategy within data warehousing, source and data integration as discussed in Chap. 3 extended with conceptual modeling of aggregation as discussed in Chap. 5, in order to illustrate the refinement of models as well as the interplay between the different perspectives in our approach. This is a part of the quality-driven DWQ data warehouse design methodology described in detail in Chap. 8.

In the context of Fig. 7.5, the example is concerned with the enterprise and source models at the conceptual perspective, with the source schemas (and possibly DW schemas) in the logical perspective, and with the extension of both to the handling of multidimensional data.

Conceptual Perspective. The DWQ approach dictates (a) the construction of a conceptual model for each source and (b) the construction of a single enterprise model that acts as the central reference model for all the other models. These models rely on an extended entity-relationship (ER) model in which both the entities and the relationships can be interpreted as concepts formalized in a description logic, and additional logical assertions can be formulated to express generic domain knowledge (*DomainAssertions*), properties and limitations of a source

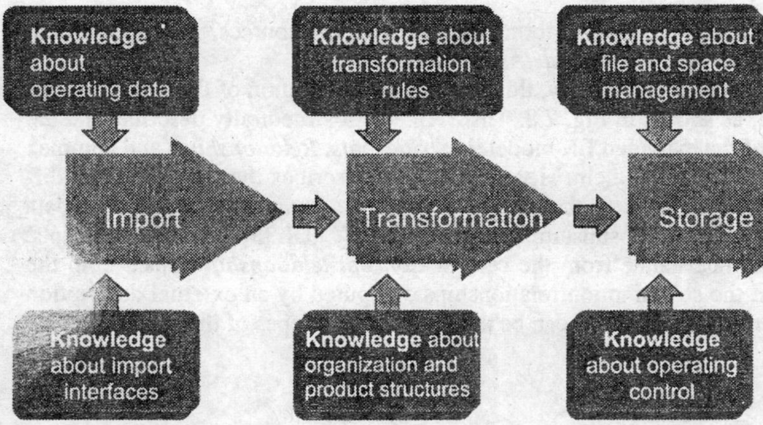

(a) Loading and refreshing the Operational Data Store

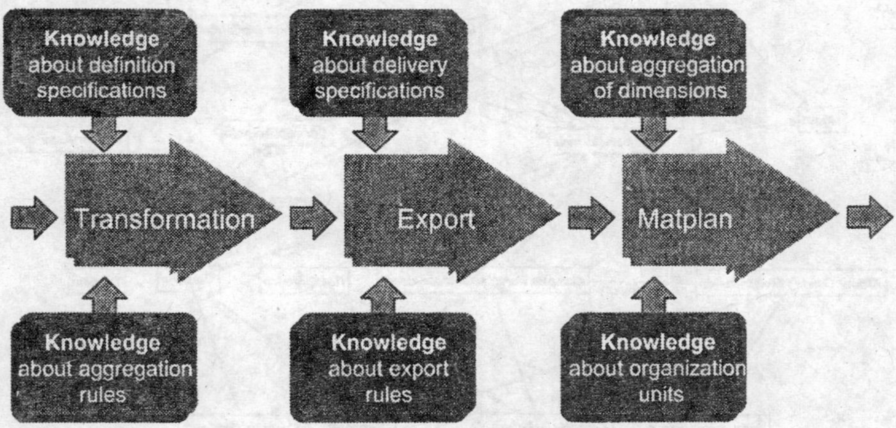

(b) Preparing a large number of specialized multidimensional models

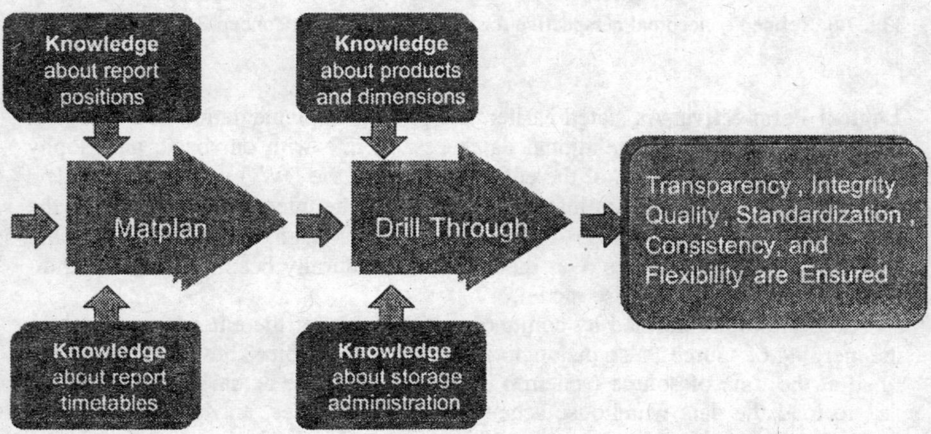

(c) Personalizing information delivery and providing access to background information

Fig. 7.7. Metadata in the data warehouse process of a large international bank [ScBJ00]

(*IntraModelAssertions*), and relationships between the sources, such as containment, consistency, etc. (*InterModelAssertions*).

In the ConceptBase repository, this leads to an elaboration of the *Concept* node from Fig. 7.5, as shown in Fig. 7.8. This refinement structurally describes the basic structure of the extended ER model, i.e., *Concepts*, *Relationships*, and complex objects constructed from them. However, it also describes the linkage of the different kinds of assertions to the objects. Despite its expressive power, this data model allows decidable subsumption reasoning [CaDL95] between concepts. Thus, through inheritance from the central *ConceptRelationship* object, both the assertions and the subsumption relationships computed by an external description logic reasoner on this structure can be applied to all subtypes of the meta schema.

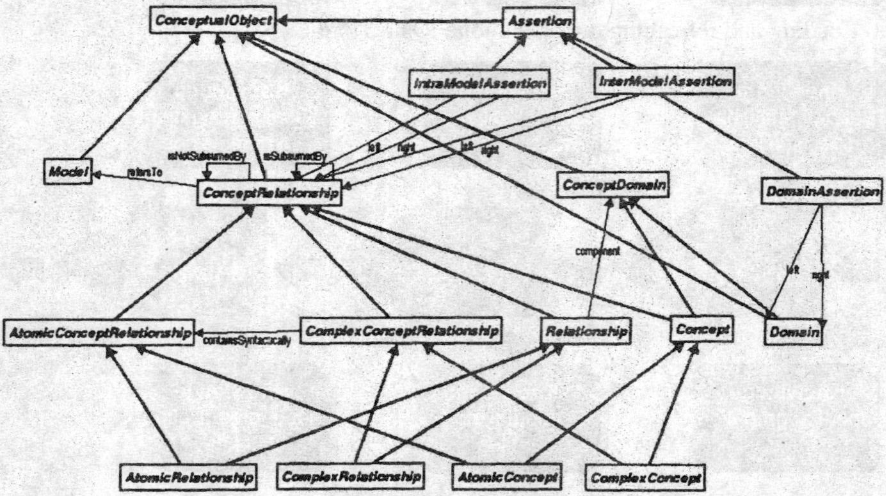

Fig. 7.8. Refined conceptual perspective for source integration (ConceptBase screendump)

Logical Perspective. As stated earlier, the present implementation of the logical perspective is limited to relational databases. In line with our basic philosophy concerning the central role of the enterprise model, the DWQ approach considers the (relational) schema of an information source to be integrated as a view on the conceptual enterprise model. As the DW schema itself consists of (possibly cleaned and merged) views over the sources, it naturally becomes also an (indirect) view over the enterprise model.

These views are defined by conjunctive *Queries* over the enterprise model. In the merging of sources, also disjunctive queries are possible. These queries are defined at the time of source (schema) integration. For the actual data integration, i.e., to load the data warehouse schema from the sources, an *AcquistionPlan* is constructed from these queries, taking into account the physical perspective. However, to capture the semantics correctly, the assertions of the conceptual model must be checked; this is accomplished by adding them as adornments to the view definition queries. From the acquisition plan and the *AdornedQueries*, an auto-

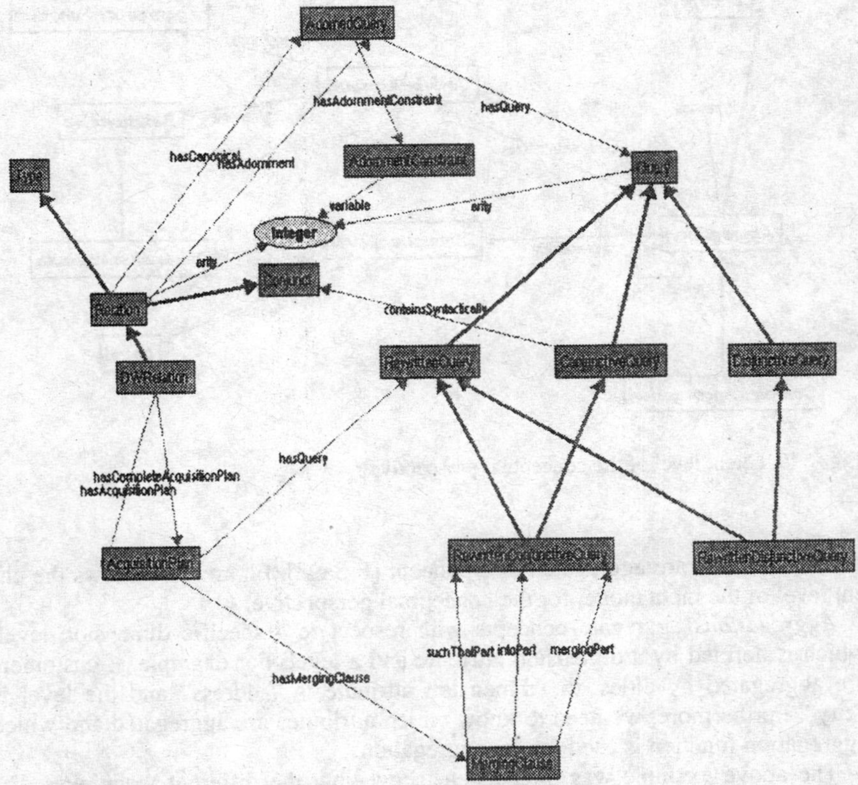

Fig. 7.9. Refined logical perspective for (relational) source and data integration

matic query rewriting defines the extraction queries from the sources as well as the *MergingClauses* that need to be executed when data from more than one source need to be merged into a data warehouse relation.

Figure 7.9 shows how this approach is captured quite naturally in the ConceptBase repository, refining the *Type* object in Fig. 7.5. This structure also provides a suitable memory for the integration process, thus allowing reuse of specific integration techniques as well as reloading of the DW. Of course, the latter is usually done incrementally by view maintenance techniques described in [StJa00].

The DWQ source and data integration approach is described in more detail in [CDL*01]. A validation case study involving the integration of four complex Telecom databases demonstrates that this information structure is suitable for the incremental modeling of data warehouse architectures; "incremental" is meant here both in the sense of gradually refining the models of a specific information source or the enterprise as a whole and in the sense of adding a new information source, possibly overlapping in concepts with the existing enterprise model.

Extension to multidimensional aggregation. The conceptual model is not restricted for the use in source integration. We can specialize the metamodel to handle also the client side of a data warehouse, i.e., multidimensional data models. In the conceptual client model, it is important how aggregations are defined and

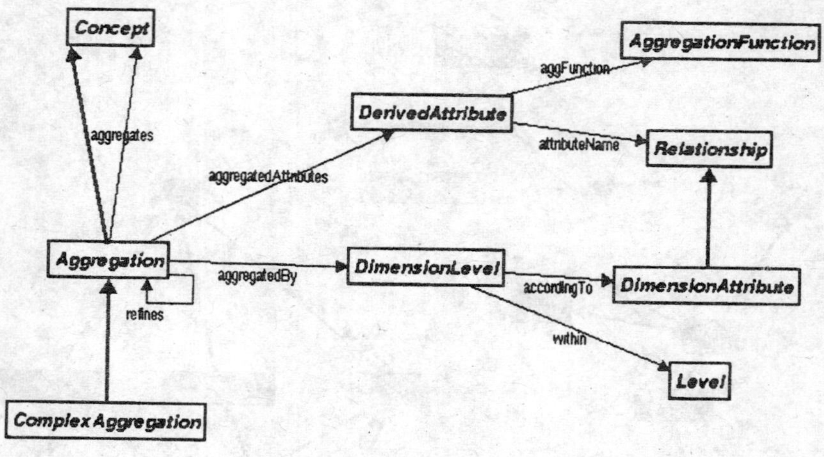

Fig. 7.10. Client level of the conceptual perspective

which attributes are aggregated of a concept [FrSa98]. Figure 7.10 shows the client level of the meta model for the conceptual perspective.

Aggregations aggregate concepts with respect to a specific dimension level, which is defined by a dimension attribute and a level. For example, if customers are aggregated by cities, the dimension attribute is "address" and the level is "city." Furthermore, we need to know which attributes are aggregated and which aggregation function is used for the aggregation.

The above example was intended to show how the different techniques discussed throughout this book can be uniformly represented via metadata under the DWQ conceptual schema. Having established this basic prerequisite, we now turn to the modeling of data warehouse quality.

7.3 Defining Data Warehouse Quality

In this section and the next one, we discuss how to extend the DW architecture model by explicit quality models and their support. There are two basic issues to be resolved. On the one hand, quality is a subjective phenomenon so we must organize quality goals according to the stakeholder groups that pursue these goals, taking into account research results in data and software quality. On the other hand, quality goals are highly diverse in nature. They can be neither assessed nor achieved directly but require complex measurement, prediction, and design techniques, often in the form of an interactive process. The overall problem of introducing quality models in metadata is therefore to achieve breadth of coverage without giving up the detailed knowledge available for certain criteria. Only if this combination is achieved, does systematic quality management become possible.

7.3.1 Data Quality

Information plays a major role in social and financial life. Organizations, governments, and companies accumulate and store information in order to process it and take advantage of it. Unfortunately, neither the accumulation nor the storage processes seem to be completely credible. In [WaRK95] it is suggested that errors in databases have been reported to be up in the 10% and even higher in a variety of applications. It is obvious that inaccurate, invalid, out-of-date, or incomplete data may have a heavy financial or social impact. Although the implementation of mechanisms for achieving data quality has financial risks and may prove not to be profitable for the organization that decides to undertake the task of implementing it, it is at least accepted that there can be an equilibrium in the cost-quality tradeoff. In [WaSF95], it is reported that more than $2 billion of U.S. federal loan money had been lost because of poor data quality at a single agency. It also reported that manufacturing companies spent over 25% of their sales on wasteful practices; the number rose to 40% for service companies.

Consequently, the problem arises on how to design and construct data warehouses that satisfy specific, well-defined quality criteria. The basic issues raised in this context are the need for a methodology for quality design for data warehouses and the modeling and the measurement of the quality of the data warehouse.

Models and tools for data warehouse quality can build on substantial previous work in the fields of data quality. Wang et al. [WaSF95] present a framework for data quality based on the ISO 9000 standard (see Appendix A – ISO 9000 for details). According to [WaSF95], *data quality policy* is the overall intention and direction of an organization with respect to issues concerning the quality of data products. *Data quality management* is the management function that determines and implements the data quality policy. A *data quality system* encompasses the organizational structure, responsibilities, procedures, processes, and resources for implementing data quality management. *Data quality control* is a set of operational techniques and activities which are used to attain the quality required for a data product. *Data quality assurance* includes all the planned and systematic actions necessary to provide adequate confidence that a data product will satisfy a given set of quality requirements.

In [WaSF95], research regarding the design of "data manufacturing systems" that incorporate data quality aspects is classified into two approaches: the development of analytical models and the design of system technologies to ensure the quality of data.

Analytical models investigate the quality in existing systems [BaPa85, BaPa87] by producing, e.g., expressions for the magnitude of errors for selected terminal outputs. One can also predict the impact of quality control efforts on such rates. In [BWPT93] a data manufacturing model to determine data product quality is proposed. The model is used for the assessment of the impact quality has on data delivered to "data customers." A "data manufacturing analysis matrix" is used to relate data units to various system components.

In the design of system technologies to ensure the quality of data, data tracking techniques have been proposed [HKRW90, PaRe90, Redm92] which use a combination of statistical control and manual identification of errors and their sources. The basis of the proposed methodology is the assumption that processes that create data are often highly redundant. The aim of the methodology is to identify pairs of steps in the overall process that produce inconsistent data.

The attribute-based model in [WaKM93, WaRG93, WaRK95, WaMa90] assumes that the quality design of an information system can be incorporated in the overall design of the system. The model proposes the extension of the relational model as well as the annotation of the results of a query with the appropriate quality indicators. Further work on data quality can be found in [BaTa89, BWPT93, Jans88, HMM*78, Krie79, AgAh87].

7.3.2 Stakeholders and Goals in Data Warehouse Quality

In order to define systematic quality management for data warehousing, we first have to identify and structure the quality goals.

There is a great deal of work related to the definition of data quality dimensions. In [BaPa82, BaPa85, BaPa87, BaTa89, BWPT93], the following quality dimensions are defined accuracy (conformity of the stored with the actual value), timeliness (recorded value is not out of date), completeness (no information is missing), and consistency (the representation of data is uniform). In [StLW94] data quality is modeled through the definition of intrinsic, contextual, representation, and accessibility aspects of data. In [Jans88, WaRG93] validation, availability, traceability, and credibility are introduced.

In software engineering, several goal hierarchies of quality factors have been proposed, including the GE Model [McRW78] and [Boeh89]. ISO 9126 [ISO91] suggests six basic factors that are further refined to an overall 21 quality factors. In [HyRo96], a comparative presentation of these three models is offered and the SATC software quality model is proposed, along with metrics for all their software quality dimensions. A structured overview of the issues and strategies, embedded in a repository framework, can be found in [JaPo92].

In [WaSF95], it is suggested that the establishment of data quality dimensions can be systematically achieved in two possible ways. The first is the use of a scientifically grounded approach in order to achieve a rigorous definition, e.g., based on information theory [DeMc92], marketing research [WaSG94], or ontology theories [WaWa96]. The second way to establish data quality dimensions is the use of pragmatic approaches, e.g., data quality dimensions can be considered as user defined.

In the following, we pursue a mixture of both approaches. Using the above-mentioned quality factors proposed in data and software engineering, we link them to the main groups of stakeholders involved in data warehouse projects, thus deriving prototypical goal hierarchies for each of these user roles.

The *Decision Maker* usually employs an OLAP query tool to get answers of interest. A decision maker is usually concerned with the *quality of the stored data*, their *timeliness* and the *ease of querying* them through the OLAP tools. The *Data Warehouse Administrator* needs facilities such as *error reporting, metadata accessibility,* and knowledge of the *timeliness* of the data, in order to detect changes and reasons for them or problems in the stored information. The *Data Warehouse Designer* needs to measure the *quality of the schemata* of the data warehouse environment (existing or newly produced) and the *quality of the metadata* as well. Furthermore, the data warehouse designer needs *software evaluation standards* to test the software packages that are being considered for purchase. The *Programmers of Data Warehouse Components* can make good use of *software implementation standards* in order to accomplish and evaluate their work. *Metadata reporting* can also facilitate their job, because they can avoid mistakes related to schema information.

Based on this analysis, we can safely argue that different roles imply a different collection of quality dimensions, which a quality model should be able to address in a consistent and meaningful way. In the following, we summarize the quality dimensions of three stakeholders, the data warehouse administrator, the programmer, and the decision maker. More details can be found in [VaBQ00].

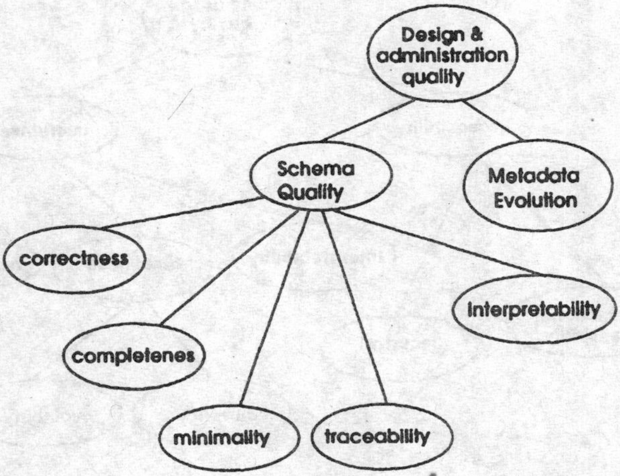

Fig. 7.11. Design and administration quality dimensions

Design and administration quality. The design and administration quality can be analyzed into more detailed dimensions, as depicted in Fig. 7.11. The *schema quality* refers to the ability of a schema or model to represent the information adequately and efficiently. The *correctness* dimension is concerned with the proper comprehension of the entities of the real world, the schemata of the sources (models), and the user needs. The *completeness* dimension is concerned with the preservation of all the crucial knowledge in the data warehouse schema (model). The *minimality* dimension describes the degree to which undesired redundancy is avoided during the source integration process. The *traceability* dimension is concerned with the fact that all requirements of users, designers, administrators and managers should be traceable to the data warehouse schema. The *interpretability* dimension ensures that all components of the data warehouse are well-described to be administered easily. The *metadata evolution* dimension is concerned with the way the schema evolves during the data warehouse operation.

Software implementation quality. Software implementation and/or evaluation is not a task with specific data warehouse characteristics. We are not actually going to propose a new model for this task but adopt the ISO 9126 standard [ISO92]. The quality dimensions of ISO 9126 are *functionality* (suitability, accuracy, interoperability, compliance, security), *reliability* (maturity, fault tolerance, recoverability), *usability* (understandability, learnability, operability), *software efficiency* (time behavior, resource behavior), *maintainability* (analyzability, changeability, stability, testability), *portability* (adaptability, installability, conformance, replaceability).

Data usage quality. Since databases and – in our case – data warehouses are built in order to be queried, the most basic process of the warehouse is the usage and querying of its data. Figure 7.12 shows the hierarchy of quality dimensions related to data usage.

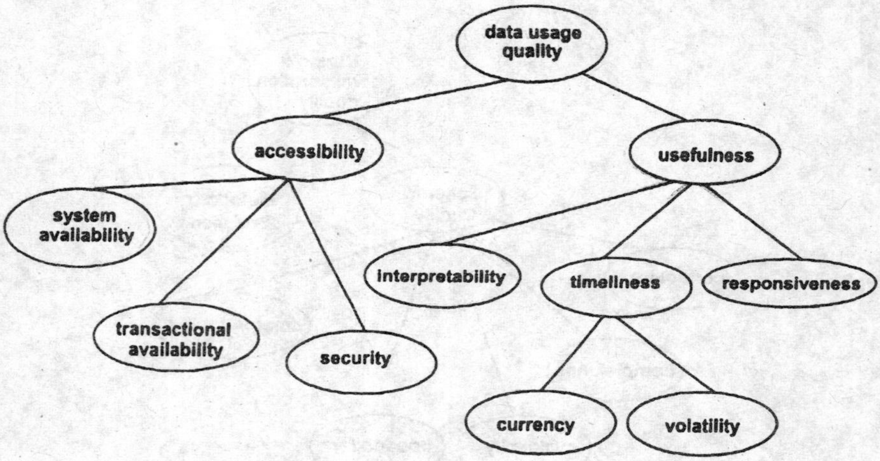

Fig. 7.12. Dimensions of data usage quality

The *accessibility* dimension is related to the possibility of accessing the data for querying. The *security* dimension describes the authorization policy and the privileges each user has for the querying of the data. *System availability* describes the percentage of time the source or data warehouse system is available (i.e., the system is up and no backups take place, etc.). The *transactional availability* dimension, as already mentioned, describes the percentage of time the information in the warehouse or the source is available due to the absence of update processes which write-lock the data.

The *usefulness* dimension describes the temporal characteristics (*timeliness*) of the data as well as the *responsiveness* of the system. The *responsiveness* is concerned with the interaction of a process with the user (e.g., a query tool which is self reporting on the time a query might take to be answered). The *currency* dimension describes when the information was entered in the sources or/and the data warehouse. The *volatility* dimension describes the time period for which the information is valid in the real world. The *interpretability* dimension, as already mentioned, describes the extent to which the data warehouse is modeled effectively in the information repository, including the question of *data lineage* (i.e., where the data come from). The better the explanation is, the easier the queries can be posed.

7.3.3 State of Practice in Data Warehouse Quality

Most (if not all) data warehouse tools affect in some way the quality of the resulting data warehouse, but only few of them directly deal with data quality. The quality of data in a data warehouse is obviously affected by three factors:

- Data warehouse schema design
- Quality of the data inserted in the data warehouse
- Manipulation of data in the data warehouse

Each of those factors is dependent on the set of tools which are used for a particular data warehouse.

Data warehouse schema design. The design of the data warehouse schema is responsible for the (semantically) correct, complete, and meaningful integration of the sources. If the design process fails to include all the required information in the data warehouse schema then the data may be ambiguous or even incomplete. If the semantics of the source data is misinterpreted or if the various sources are not properly integrated then the data warehouse will contain incorrect data. Also if the design process does not identify the required integrity constrains, the data warehouse may store meaningless or incorrect information. All the quality dimensions of [WaWa96]–completeness, unambiguousness, meaningfulness and correctness–are affected by the design process.

The design of the data warehouse schema is a complicated process involving the analysis of requirements, analysis of the available data, schema extraction and integration (of the sources), and other general database design steps. The tools which may assist in this process belong to the following categories:

- CASE tools
- Data modeling
- Database design
- Schema integration
- Metadata management
- Data reverse engineering

Quality of the inserted data. Obviously the data stored in the data warehouse depends on the quality of data used to load/update the data warehouse. Incorrect information, stored at the data sources, may be propagated in the data warehouse. Still, data are inserted in the data warehouse through a load/update process which may (or may not) affect the quality of the inserted data. The process must correctly integrate the data sources and filter out all data that violate the constrains defined in the data warehouse. The process may also be used to further check the correctness of source data and improve their quality. The tools which may be used to extract/transform/clean the source data or to measure/control the quality of the inserted data can be grouped in the following categories [Orli97]:

- Data extraction
- Data transformation
- Data migration
- Data scrubbing
- Data cleaning

- Data content quality
- Data quality analysis

Manipulation of data. Data in a data warehouse are usually handled by a Database Management System (DBMS) and cannot be updated by users. The most common manipulations are aggregations and multidimensional data reorganization which are carried out by the DBMS. This means that the quality of data is generally preserved inside the data warehouse, and it is hardly affected by the manipulation processes. In most cases the only tools used to manipulate the data in the data warehouse belong to the following categories:

- General purpose database management systems
- Multidimensional database management systems

Products. The major database/software vendors (IBM, Oracle, Sybase, Microsoft, Tandem) provide quality-oriented tools that cover nearly all the previously mentioned categories. Smaller vendors have produced a large variety of specialized tools. Those tools mainly belong to the categories:

- Data extraction
- Data transformation
- Data migration
- Data scrubbing
- Data cleaning
- Data content quality
- Data quality analysis
- Schema integration
- Metadata management
- Data reverse engineering

Despite all of these tools, the quality-oriented design and management of a data warehouse remains a major challenge for which few comprehensive methodologies exist. In the sequel, some promising approaches will be described.

7.4 Representing and Analyzing Data Warehouse Quality

We now turn to the formal handling and repository-based management of DW quality goals such as the ones described in the previous section. First, we discuss the quality function deployment (QFD) approach as a method for quality planning, then the goal-question-metric (GQM) approach as a method for quality evaluation in data warehousing. Subsequently, we continue our description of some of the DWQ solutions by extending the architecture metadata described in Sect. 7.2 by explicit repository support of quality management.

7.4.1 Quality Function Deployment

A first formalization could be based on a qualitative analysis of relationships between the quality factors themselves, e.g., positive or negative goal-subgoal relationships or goal-means relationships. The stakeholders could then enter their subjective evaluation of individual goals as well as possible weightings of goals and be supported in identifying good trade-offs. The entered as well as computed evaluations could be used as quality measurements in the architecture model of Fig. 7.4, thus, enabling a simple integration of architecture and quality model.

Such an approach is widely used in industrial engineering under the label of QFD, using a special kind of matrix representation called the *House of Quality* [Akao90], see Fig. 7.13. Formal reasoning in such a structure has been investigated in works about the handling of nonfunctional requirements in software engineering, e.g., [MyCN92]. Visual tools have shown a potential for negotiation support under multiple quality criteria [GeJJ97].

The methodology for building a house of quality comprises several steps. The first step involves the modeling of customer needs and expectations. This step produces a list of goals-objectives, often referred as the "WHATs" [BBBB95]. It is very possible that a customer requirement is expressed rather generally and vaguely; so the initial list is refined and a second, more detailed, list of customer requirements is produced. If it is necessary, this procedure is repeated.

The second step involves the suggestion of technical solutions (the "HOWs") which can deal with the problem that was specified at the previous step. This process can also be iterated, as it is rather hard to express detailed technical solutions at once.

The third step involves the combination of the results of the two previous steps. The basic aim of this process is to answer the question "how are customer requirements and possible technical solutions interrelated?" To achieve that, the interior of the house of quality, called the *relationship matrix*, is filled in. Symbols are usually used, determining how strong a relationship is. It is also very important to note that both positive and negative relationships exist.

The fourth step involves the identification of interrelationships between the technical factors. The roof of the house, known as the *correlation matrix* is filled in. All the conflicting points represent trade-offs in the overall technical solution.

Next, *competitive assessments* must be made. They comprise a pair of weighted tables which depict analytically how competitive products compare with the organization products. The competitive assessment is separated in two categories: customer assessment and technical assessment.

The following step is the prioritization of customer requirements. The prioritized customer requirements are a block of columns corresponding to each customer requirement and contain columns for importance rating, target value, scale-up factor, sales point, and absolute weight for each customer requirement.

Finally, prioritized technical descriptors are also defined. Each of the proposed technical solutions is annotated with degree of technical difficulty, target value, and absolute and relative weights.

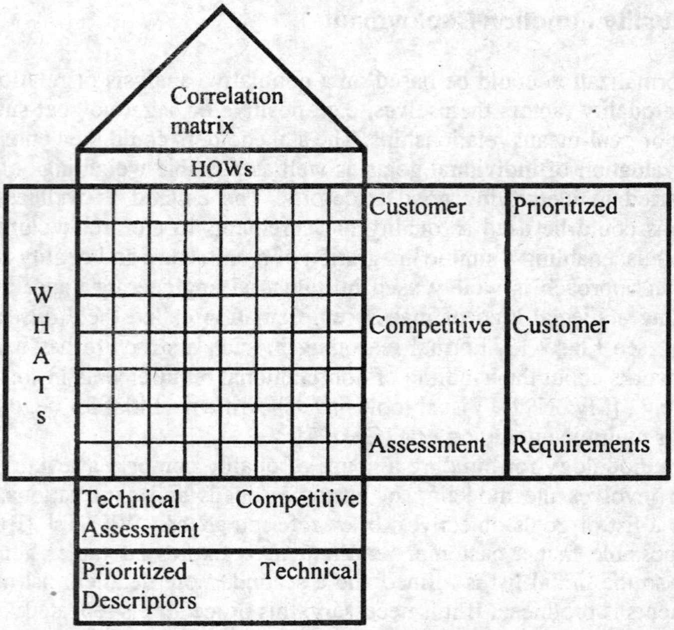

Fig. 7.13. The House of Quality in Quality Function Deployment

7.4.2 The Need for Richer Quality Models: An Example

While QFD certainly has a useful role in rough quality planning and cross-criteria decision making, using it alone would throw away the richness of work created by research in measuring, predicting, or optimizing individual DW quality factors. Such methods need to be systematically adopted or newly developed for all quality factors found important in the literature. To give an impression of the richness of techniques to be considered, we use a single quality factor – responsiveness in the sense of good query performance – to illustrate three complementary approaches, one each from the conceptual, logical, and physical perspective.

We start with the logical perspective [ThSe97] which is used as a detailed example in Sect. 7.5. Here, the quality indicator associated with responsiveness is taken to be a weighted average of query and update "costs" for a given query mix and given information sources. A combinatorial optimization technique selects a collection of materialized views as to minimize the total costs. This can be considered a simple case of the above QFD approach, but with the advantage of automated design of a solution.

If we include the physical perspective, the definition of query and update "costs" becomes an issue in itself: what do we mean by costs – response time, throughput, or a combination of both (e.g., minimize query response time and maximize update throughput), what actually produces these costs – is database access or the network traffic the bottleneck. A comprehensive queuing model [NiJa99] enables the prediction of such detailed metrics from which the designer can choose the right ones as quality measurements for the design process. In addition, completely new design options come into play: instead of materializing more views to improve query response time (at the cost of disturbing the OLTP updates), the designer could buy a faster client PC or DBMS, or use an ISDN link rather than using slow modems.

Yet other options come into play if a rich logic is available for the conceptual perspective. The description logic used in the DWQ approach [CDL*97] allows to state that, for example, information about all instances of one concept in the enterprise model is maintained in a particular information source, i.e., the source is complete with respect to the domain. This enables the DW designer to drop the materialization of all views on other sources, thus reducing the update effort semantically without any loss in completeness of the answers.

7.4.3 The Goal-Question-Metric Approach

It is clear that there can be no decidable formal framework that even comes close to covering all of these aspects in a uniform language. When designing the metadatabase extensions for quality management, one therefore needs to look for semiautomatic solutions that still maintain the overall picture offered by the shallow quality management techniques discussed at the beginning of this section but are at the same time open for the embedding of specialized techniques.

Our solution to this problem builds on the widely used GQM approach to software quality management [OiBa92]. The idea of GQM is that quality goals can usually not be assessed directly, but their meaning is circumscribed by questions that need to be answered when evaluating the quality. Such questions again can usually not be answered directly but rely on metrics applied to either the product or process in question; techniques such as statistical process control charts are then applied to derive the answer of a question from the measurements.

The GQM process consists of the following steps:

1. *Identify quality and/or productivity goals* at corporate, division or project level; e.g. customer satisfaction, on-time delivery, improved performance;
2. From those goals and based upon models of the object of measurement, derive *questions that define those goals* as completely as possible;
3. *Specif the measures* to be collected in order to answer those questions;
4. *Develop data collection mechanisms*, including validation and analysis mechanisms.

A *goal* is defined with respect to an *issue* (e.g., timeliness), an *object* (e.g., change request processing), a *viewpoint* (e.g., project manager) and a *purpose* (e.g., improvement). The *issue* and the *purpose* of the goal are obtained from the *policy and the strategy of the organization* (e.g., by analyzing corporate policy statements, strategic plans and, more importantly, interviewing relevant subjects in the organization). The *object* coordinate of the goal is obtained from a description of the process and products of the organization, *by specifying process and product models*, at the best possible level of formality. The *viewpoint* coordinate of the goal is obtained from the *model of the organization*.

There are three types of questions:

Group 1. How can we characterize the object (product, process, or resource) with respect to the overall goal of the specific GQM model? For instance,

- What is the current change request processing speed?
- Is the change request process actually performed?

Group 2. How can we characterize the attributes of the object that are relevant with respect to the issue of the specific GQM model? For example,

- What is the deviation of the actual change request processing time from the estimated one?
- Is the performance of the process improving?

Group 3. How do we evaluate the characteristics of the object that are relevant with respect to the issue of the specific GQM model? For example,

- Is the current performance satisfactory from the viewpoint of the project manager?
- Is the performance visibly improving?

In each major data warehouse development and management effort, the development of quality metrics is a customized process which should target for the *maximization of the use of existing data sources*, the application of *objective* measures to more *mature* measurement objects, and more *subjective* evaluations to *informal or unstable* objects.

7.4.4 Repository Support for the GQM Approach

We now return to the repository model of the extended DW architecture presented in Section 7.2, and show how explicit quality management can be integrated into it, using a metamodel of the GQM approach. This approach goes beyond that of other GQM tools because the query mechanisms of the knowledge-based repository can be exploited.

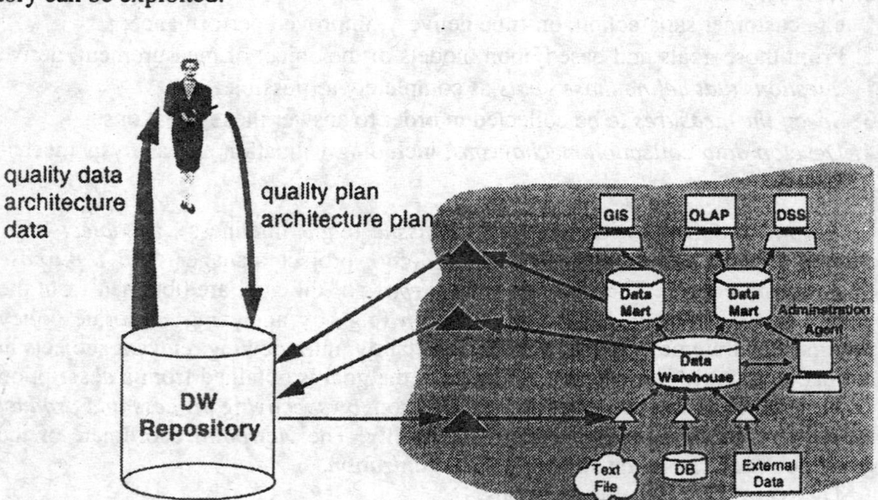

Fig. 7.14. Quality management via the data warehouse repository

Our GQM version is defined through the idea of quality queries as materialized views over the data warehouse; the views are defined through generic queries over the quality measurements. Figure 7.13 motivates this approach by zooming in on the repository. The stakeholder assesses the data warehouse quality by asking quality queries to the repository. The repository answers the queries by accessing quality data obtained from measurement agents (the black triangles in Fig. 7.13). The agents communicate with the components of the real data warehouse to extract measurements.

The stakeholder may redefine the quality goals at any time. This shall lead to an update of the quality model in the repository and possibly to the configuration of new measurement agents responsible to deliver the base quality data. Analogously, a stakeholder with appropriate authorization can redefine the architecture of the data warehouse via the repository. Such an evolutionary update, e.g., the specification of a new data source, leads to a reconfiguration of the real data warehouse. Ultimately, the quality measurements will reflect such effect of the change and give evidence whether the evolution has led to an improvement of some quality goals.

The use of the repository for data warehouse quality management has significant advantages:

- Data warehouse systems already incorporate repositories to manage metadata about the data warehouse; extending this component for quality management is a natural step.
- Existing metadata about the data warehouse, e.g., source schemas, can be directly used for formulating quality goals and measurement plans.
- The quality model can be held consistent with the architecture model, i.e., the repository can prevent the stakeholders to formulate quality goals that cannot be validated with the given architectural data.
- The stakeholder accesses the repository as a *data source* to deliver quality reports to the stakeholders who formulate quality goals; in fact, producing such reports is the same kind of activity that is used to deliver aggregated data to the client tools of a data warehouse.

The last argument is not just a technical remark. Quality data, i.e., values of quality measurements, are derived from DW components. The values are materialized views of properties of these components. These values do have quality properties like timeliness and accuracy themselves. It makes a difference whether the value of a quality measurement is updated each hour or once a month. While we do not go into detail with this "second-level" quality, we note that the same methods that are used to maintain quality of the DW can also be used to maintain the quality of the DW repository (hosting the quality model).

7.4.4.1 The Quality Meta Model

Quality data is derived data and is maintained by the data warehouse system. This implementation strategy provides more technical support than GQM implementations for normal software systems. Such systems lack the built-in repository. The expressive query language offered by the ConceptBase repository system makes a large portion of quality management tasks a matter of query formulation. In the sequel, we elaborate how a version of GQM can be modeled by Telos metaclasses in ConceptBase and then be used for quality goal formulation and quality analysis.

Telos provides a uniform logical representation for different object relationships, such as class membership *(x in class)*, specialization between classes *(c isA d)*, and attributes *(x label y)*. This logical representation can be mapped to a graphical layout as shown for the quality model below, as well as to a frame syntax which we sometimes use for the formulation of queries. Since all items (objects, classes, metaclasses, and attributes) are uniformly treated in the logical representation, the Telos language is used for formulating.

1. A metamodel by a collection of metaclasses (here for defining the architecture and quality models);
2. A collection of classes (here the use of the architecture and quality metamodels to express quality goals, queries, and measurement types on DW components;
3. Instances of the classes (here for representing results of measurements as class instances).

Figure 7.14 shows the Telos metaclasses for managing data warehouse quality. Quality goals, e.g., "improve the timeliness of data set sales-per-month," are assigned to stakeholders. The *purpose* attribute for quality goals is used to specify the intended direction of quality improvement (e.g., to increase the quality or to achieve a certain quality level at a certain time). The quality goal is imposed on

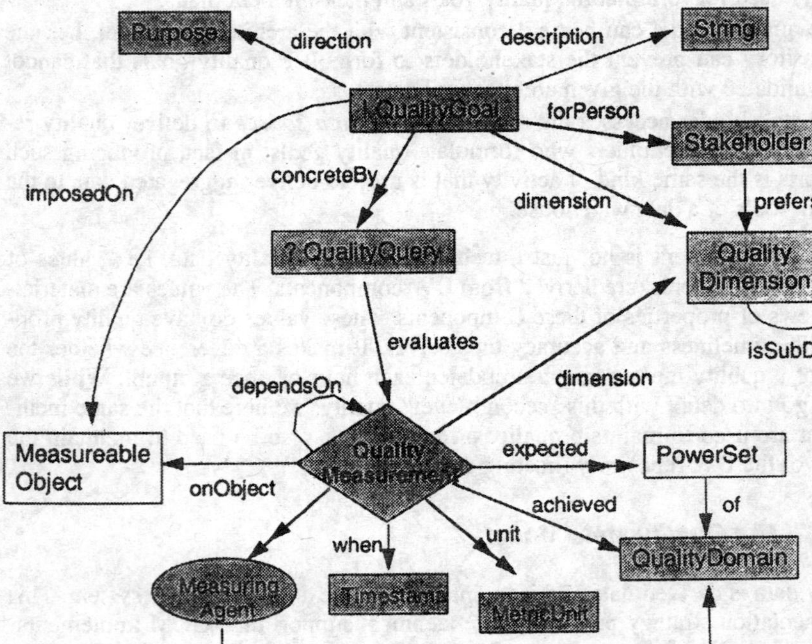

Fig. 7.15. A metamodel for data warehouse quality

measurable data warehouse objects as classified by the architecture model of Fig. 7.4. A quality goal is linked to one or more *quality dimensions* according to the preferences of the stakeholder who formulates the goal (see Sect. 7.3.1).

Quality goals are mapped to a collection of *quality queries* which are used to decide whether a goal is achieved or not. While questions are just textual in the original GQM approach, we encode a quality query as an executable query on the data warehouse repository using the expressive deductive query language of ConceptBase. The answer to a quality query is regarded as evidence for the fulfillment of a quality goal. The simplest kind of quality query would just evaluate whether the current *quality measurement* for a data warehouse object is within the expected interval. A quality measurement uses a metric unit, e.g., the average number of null values per tuple of a relation.

7.4.4.2 Implementation Support for the Quality Metamodel

The abstraction levels of the concepts in the quality model require a closer consideration [JeQJ98]. In standard software metrics, a *quality measurement* is a function that maps a real-world entity to a value of a domain, usually a number. In our case, we maintain abstract representations of all "interesting" real-world entities in the DW repository itself. Thus, quality measurements can be recorded as explicit relationships between the abstract representations, i.e., measurable objects, and the quality values. By nature, such a quality measurement relates objects of different abstraction levels. For example, a quality value of 0.8 could be measured for the percentage of null values of the *Employee* relation of some data source. *Employee* is a relation whereas 0.8 is just a number. For this reason, we require a framework such as Telos which is able to relate objects at different abstraction levels.

A second remark concerns the use of the quality model by instantiation. Typical instances of *MeasurableObject* are items like *relation* (logical perspective) or *entity type* (conceptual perspective). These items are independent of the DW application domain. They are used to describe a DW architecture but they are not components of a concrete DW architecture[1]. A concrete architecture consists of items like data source for *Employee*, concrete wrapper agents, etc. Therefore, when we instantiate the quality model, we describe types of quality goals, types of queries, and types of measurements. For example, we can describe a completeness goal for relational data sources (instances of the *Relation* concept in Fig. 7.9) which is measured by counting the percentage of null values in the relation. Such types (or patterns) can be reused for any concrete DW architecture. For example, the measurement for a relational source for *Employee* would be instantiated from the measurement type by instantiating the expected and achieved quality values. This two-step instantiation is essential since it allows preloading the repository with quality goal, query, and measurement types independent of the application domain.

Quality goals – whose dimensions are organized in hierarchies such as shown in Figs. 7.11 and 7.12 – are made operational as types of queries defined over

[1] Formally, this is expressed by means of class instantiation in Telos. The concept *Relation* is represented by a tuple *(Relation in MeasurableObject)*. The concept *Employee* is introduced in Telos by a tuple *(Employee in Relation)*. Thus, *MeasurableObject* is a metaclass of *Employee*.

quality measurements. These queries will support the evaluation of a specific quality goal when parameterized with a given (part of a) DW metadatabase. Such a query usually compares the analysis goal to a certain expected interval in order to assess the level of quality achieved.

As a consequence, the quality measurement must contain information about both expected and actual values. Both could be entered into the metadatabase manually or computed inductively by a given metric through a specific reasoning mechanism. For example, given a physical design and some basic measurements of component and network speeds, the queuing model in [NiJa99] computes the quality measurement response time and throughput, and it could indicate if network or database access is the bottleneck in the given setting. This could then be combined with conceptual or logical quality measurements at the level of optimizing the underlying quality goal.

Generally speaking, quality queries access information recorded by quality measurements. A quality measurement stores the following information about data warehouse components:

1. An interval of expected values
2. The achieved quality measurement
3. The metric used to compute a measurement
4. Causal dependencies to other quality measurements

The dependencies between quality measurements can be used to trace quality problems, i.e., measurements that are outside the expected interval, to their causes. The following ConceptBase queries exemplify how quality measurements classify data warehouse components and how the backtracing of quality problems can be done by queries to the metadata base:

```
QualityQuery BadQuality isA QualityMeasurement
with constraint
    c: $ not (this.expected contains this.current) $
end
QualityQuery CauseOfBadQuality isA DW_Object
with parameter
  badObject : DW_Object
constraint
  c: $ exists q1,q2/QualityMeasurement
     (badObject classifiedBy q1) and (q1 in BadQuality) and
     (q1 dependsOn q2) and (q2 in BadQuality) and
     ((this classifiedBy q2) or  (exists o/DW_Object (o classifiedBy q2) and
       (this in CauseOfBadQuality[o/badObject]))) $
end
```

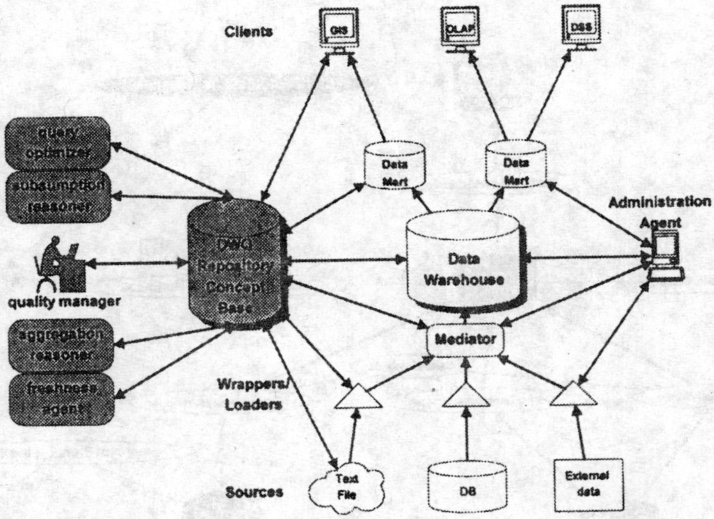

Fig. 7.16. Mapping the DWQ metamodel to the traditional data warehouse architecture

7.4.4.3 Controlling and Improving Quality with the Repository

Summarizing the discussion above, Fig. 7.16 gives an impression how the traditional data warehouse architecture is extended by our repository centered metadata management approach. The quality model forms the basis of the implementation in ConceptBase. Quality data (i.e., values of measurements) are entered into the ConceptBase system by external measurement agents which are specialized analysis and optimization tools. In the DWQ project, four such tools have been developed. Besides the subsumption reasoning tools already mentioned in Sect. 7.2.6, they include a data freshness toolkit covering the physical modeling of source integration, and tools for reasoning about multidimensional aggregates and query optimization on the client side. ConceptBase can trigger these agents based on the timestamp associated with them in the repository.

The result of the analysis of the quality data can be displayed graphically, as illustrated in Fig. 7.17. Quality measurements are the wide ovals in the middle. The black oval indicates that the timeliness of the staff department data store (an item of the physical perspective) is not in its expected range. The white color of the other measurements indicates measurements that are in expected range. The color code of the graphical view is computed by the repository based on the *BadQuality* query shown above.

The graphical display is intended for controlling the quality of the data warehouse. The "black" nodes indicate locations where some ad hoc *control* is required or where stakeholders have to be aware of unexpected low quality. Stakeholders have their own quality goals and hence have individualized views on the quality.

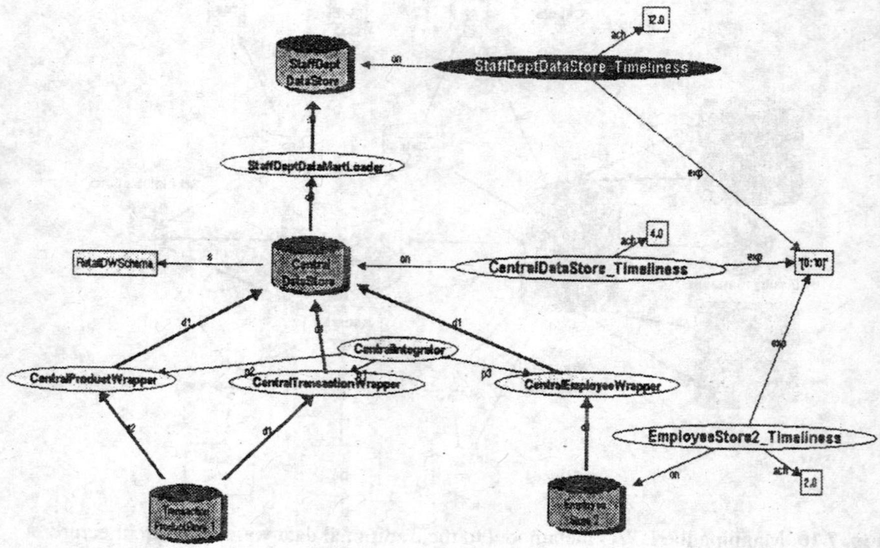

Fig. 7.17. ConceptBase screenshot of the graphical view on the quality data

The repository can also be used to maintain the knowledge about causes of quality measurements. The *dependsOn* link in Fig. 7.15 is intended to build such a symptom-to-cause model over the quality measures. Such a mathematical model shall be used to *understand* the effects of certain measures to other (dependent) measures. As soon as the mathematical models are coded into the repository, they can be used to forecast derived quality measures. If derived and measured values coincide for the same parameter, then the model is validated. This issue is still under research in the data warehouse area, however.

The last and most advanced aspect of quality management is the *improvement*. In the original GQM approach, the results of measurements may cause specific improvement actions. However, quality goals can also directly be associated with a data warehouse design process, as illustrated next.

7.5 A Detailed Example: Quality Analysis in Data Staging

To conclude this chapter, we illustrate the interplay of architecture, processes, and quality described in this chapter in more detail, using the example of quality analysis and quality-driven design of the data staging area (DSA) [VLST01].

The DSA can be considered as a set of (possibly) distributed materialized views defined over the data provided by the data sources. These views are used to answer all the queries posed to the DSA, thus avoiding going back to the original sources. On the other hand, when modifications occur to the sources, the materialized views have to be refreshed through insertions, deletions, and updates. For a given set of different source databases and a given set of queries that the data warehouse has to service, there is a number of alternative sets of materialized views that the administrator can choose to maintain. Each of these sets provides different values for the various quality aspects relevant to the DSA design. Such a quality factor is, e.g., the *DSA operational cost* which depends on the cost of two basic operations: query answering and the view refreshing. Trying to select a set of views to materialize in order to maximize the overall DSA quality is the *Quality-Oriented DSA Design Problem.*

In terms of the architecture part of metadata repository, the problem involves three basic object types: the *sources*, which are considered to be relational (or at least provide a relational interface through a gateway or wrapper), the *queries* posed from the user over these sources, and the *intermediate materialized views* that the algorithms produce, which serve as buffers speeding up the answering of the queries. Another interesting object of the problem, although implicit in the problem definition, is the employed algorithm itself.

The basic *DSA quality dimensions* are mainly related to time and space:

1. *Schema Minimality.* The redundancy in the data staging area is kept low, for ease of administration.
2. *Design Efficiency.* The choice of the materialized views must be done as quickly as possible.
3. *Design Consistency.* The constraints over the result of the design process are respected (e.g., the volume of the produced materialized views must fit in the disk space provided for the data staging area).
4. *Query Timeliness.* The answering of the queries must be done as quickly as possible.
5. *System Availability.* The maintenance of the materialized views must be done as quickly as possible.
6. *Data Quality.* The data delivered to the users must be accurate, complete, up to date, and abiding by the internal rules (constraints) of the data warehouse.

For each of the related objects of the problem, we can define relevant *quality factors* that affect the result of the design of the data staging area. The views of the DSA are related to several quality factors such as the *final number of views in the DSA*, the *available disk space for the DSA*, the *space occupied by each view* (as well as the *total space occupied by al the views*), the *update cost for view* (i.e., the time to perform the update of a view of the data staging area), as well as the *total update time for the DSA*. The queries are characterized from their *query cost* (i.e., the time needed to perform the query over the data staging area). All objects are also characterized by their data quality. The quality factors related to data quality are the *accuracy* (i.e., the validity of the data, with respect to real-world values), the *completeness* (i.e., the percentage of available information with respect to expected volume of information), the *consistency* (i.e., the percentage of information obeying the database rules), and *the freshness* (i.e., the age of the data with respect to their transaction time). Finally, the design algorithm itself is characterized by the *design time* (i.e., the total time needed for the algorithm to terminate). All the aforementioned quality factors are listed in Table 7.1, along with their precise definition and measurement method.

The overall problem of data warehouse design optimization, composed of its objects and quality factors is graphically depicted in Fig. 7.18. For each of the basic object types (i.e., sources, DSA views, and queries), we give their relevant quality factors (as already described in this section). Apart from the quality factors and the participating objects, Fig. 7.18 shows two *quality goals*. On the left side, the quality goal "*design an DSA respecting the quality constraints set by the administrator*" is depicted. On the right side, we depict the administrator's quality goal "*evaluate the quality of the produced DSA and the respective queries*".

In the sequel, we present a detailed cost-benefit model by which design alternatives for DSA view materialization can be evaluated starting from the quality goals. In the evaluation itself, quality factors then act as thresholds for a proposed or existing combination of materialized views. For example, one can have constraints like: "the size of the result must not exceed the size of the disk," or "the completeness of all the views is greater than 80%". These problems can be dealt with trivially since each time a new state is explored, it can be approved or not by evaluating the predefined constraints over this solution.

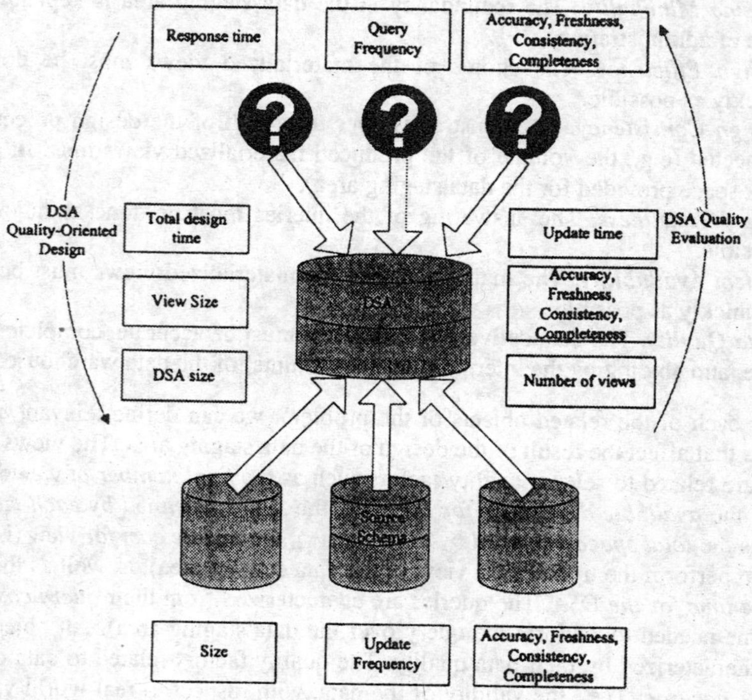

Fig. 7.18. The quality-oriented DSA design and evaluation problem [VaBQ00]

Table 7.1. Quality factors for the data warehouse design optimization problems

Quality Factor	Definition	Measurement Method	Related Dimension
Number of DSA views	The final number of views in the DSA	Simple enumeration	Schema minimality
Design time	Total time needed for the algorithm to terminate	Through a trivial counter	Design efficiency
Disk space	Available disk space for the DSA	Trivial	Design consistency
Space occupied by each view	Space occupied by each view	Rows * size(row)	
DSA space	Total space occupied from all the DSA views	Sum of the space occupied by all the views	
Query cost for view	Time to perform the query over the DSA	See cost model (Sect. 3)	Query timeliness
Total query cost	Total time needed for all the queries to be answered from the DSA	Sum of all the query costs	
Update cost for view	Time to perform the update of a DSA view	See cost model (Sect. 3)	System availability
Total update cost	Total time needed to update all the views of the DSA	Sum of all the update costs	
Accuracy	Validity of the data with respect to real-world values.	Percentage of true over stored facts or estimation or sampling	Data quality
Completeness	The percentage of available information wrt. expected volume of information	Percentage of stored over expected facts or estimation or sampling	
Consistency	Percentage of information obeying the database rules	Percentage of facts obeying rules over stored facts or estimation	
Freshness	Age of data wrt. their acceptable age	Maximum timestamp of the data obtained from log scan, or a dedicated column, or temporal DBMS techniques	

7.5.1 Evaluation of the Quality of a DSA Schema

In this subsection we formulate the problem mathematically as a state space problem and introduce a cost model for the evaluation of the benefit (or actually cost) of each state, which takes into consideration the quality of the sources.

Mathematical formulation of the problem. We consider that a nonempty set of queries **Q** is given, defined over a set of source relations **R**. The data warehouse contains a set of materialized views **V** over **R** such that every query in **Q** can be rewritten completely over **V**. Thus, all the queries in **Q** can be answered locally at the data warehouse, without accessing the source relations in **R**. By Q^V, we denote a complete rewriting of the query $Q \in \mathbf{Q}$ over V.

Consider a *Data Warehouse Configuration* (i.e., a set of views **V** and a set of rewritings of the queries over **V**, $\mathbf{Q}^V$) $C = <\mathbf{V}, \mathbf{Q}^V>$. The data warehouse design problem can then be stated as follows:

Input:

- A set of source relations **R**
- A set of queries **Q** over **R**
- A set of constraints involving specific thresholds for several quality factors
- A cost model involving different quality factors, each with different importance (weight)

Output:

- A data warehouse configuration $C = < \mathbf{V}, \mathbf{Q}^V >$ such (a) that all the constraints are satisfied and (b) the cost of the solution is minimal.

The problem is investigated for the case of select-project-join (SPJ) conjunctive queries without self-joins. The relation attributes take their values from domains of integer values. Atomic formulas are of the form $x \text{ op } y + c$ or $x \text{ op } c$, where x, y are attribute variables, c is a constant, and op is one of the comparison operators =, <, >, ≥, ≤, but not ≠. A formula F implies a formula F' if both involve the same attributes and F is more restrictive (for example A=B implies A ≤ B+10). Atoms with attributes from only one relation are called *selection atoms*, those with attributes from two relations are called *join atoms*.

A set of SPJ views **V** can be represented by a multiquery graph. A multiquery graph allows the compact representation of multiple views by merging the query graphs of each view. For a set of views **V**, the corresponding multiquery graph, G^V, is a node and edge labeled multigraph resulting by merging the identical nodes of the query graphs for the views in V.

We define the following five transformation rules that can be applied to a data warehouse configuration (see [ThSe97] for a broader discussion):

Edge Removal. A new configuration is produced through the elimination of an edge label by the atom p from the query graph of view V and the addition of an associated condition into the queries that are defined over this view. If the query graph of V was not kept connected after the edge elimination, V is decomposed in two separate views.

Attribute Removal. If we have atoms of the form A=B and A, B are attributes of a view V, we eliminate A from the projected attributes of V.

View Break: Let V be a view and N_1, N_2 two sets of nodes labeled by V in G^V such that: (a) $N_1 \not\subset N_2$, $N_2 \not\subset N_1$, (b) $N_1 \cup N_2$ is the set of all the nodes labeled by V in G^V, and (c) there is no edge labeled by V between the nodes in $N_1 - N_2$ and $N_2 - N_1$. In this case, we replace V by two views V_1 (defined over N_1) and V_2 (defined over N_2). All the queries defined over V are modified to be defined over $V_1 \bowtie V_2$.

View Merging. A merging of two views V_1 and V_2 can take place if every condition of V_1 (V_2) implies or is implied by a condition of V_2 (V_1). In the new configuration, V_1 and V_2 are replaced by a view V that is defined over the same source relations and comprising all the implied predicates. All the queries defined over V_1 or V_2 are modified appropriately in order to be defined over V. Eliminated selection and join conditions are added to these queries.

Attribute Transfer. Let us have atoms of the form A=c where A is an attribute of a view V of a labeled node N and c a constant. We replace V by V_1 with the same nodes and edges but without the attribute A. We also create the view $V_2 = \pi_A(\sigma_{A=c} N)$). We replace every occurrence of V in Q^V by $V_1 \bowtie V_2$.

Quality evaluation. We define the cost of a specific *state* as a function of three major quality factors: query cost, update cost (which combined form the composite quality factor operational cost), and data quality (which is the combination of quality factors such as accuracy, completeness, freshness, and consistency). Formally, if **V** is a state, and w, w_q, w_u, w_1, w_2, w_3, w_4, are constant positive weights, smaller or equal to 1, such that $w_q + w_u = 1$ and $w_1 + w_2 + w_3 + w_4 = 1$ we have the following formulas:

$$\text{Operational cost}(\mathbf{V}) = w_q * \text{query cost}(\mathbf{V}) + w_u * \text{update cost}(\mathbf{V})$$

$$\text{Data quality}(\mathbf{V}) = w_1 * \text{accuracy}(\mathbf{V}) + w_2 * \text{freshness}(\mathbf{V}) + w_3 * \text{completeness}(\mathbf{V}) + w_4 * \text{consistency}(\mathbf{V})$$

$$\text{Cost}(\mathbf{V}) = w * \frac{\text{operational cost}(\mathbf{V})}{\text{data quality}(\mathbf{V})}$$

To compute the quality factors of a state, we employ the quality factors of the materialized views that comprise this state. More specifically, if we denote a state **V** as a set of views {V}, we can compute its quality factors as follows:

$$\text{Query cost}(\mathbf{V}) = \Sigma_V \text{query cost}(V)$$

$$\text{Update cost}(\mathbf{V}) = \Sigma_V \text{update cost}(V)$$

$$\text{Accuracy}(\mathbf{V}) = \frac{\sum_V \text{popularity}(V) * \text{accuracy}(V)}{\sum_V \text{popularity}(V)}$$

$$\text{popularity}(V) = \text{number of queries using } V$$

The formula for accuracy applies analogously for completeness, consistency, and freshness. Note that we choose the popularity (i.e., frequency of reference) of a view as its weight in the formulas for data quality. This is not a clear choice, since other measures (e.g., size or number of rows) could be used for that purpose. as [Orr98] mentions “the problem of data quality is fundamentally intertwined in how ... users actually use the data in the system,” since the users are the ultimate judges of the quality of the data produced for them. Also, since data quality is a positive number, the greater its value the lower the respective “cost” value.

7.5.2 Analyzing the Quality of a View

After defining the quality factors for a state as·functions over the quality factors of its views, all that remains is to define the formulas for the quality factors of a view, by using the quality factors of the source relations. The evaluation of the query cost of selections, projections, and joins is done in terms of estimation of the time needed to compute the result. For the computation of the update cost, for each source table R, the model assumes the existence of two auxiliary differential tables ΔR^+, ΔR^-, which are used to capture the insertions and deletions over the previous version of R. The specific formulas are found in [ThLS99].

We provide specific cost formulas for different quality factors, namely *freshness, accuracy, completeness,* and *consistency.* Given a source relation R, we define metrics for the expected quality of $\pi_A(R)$, $\sigma_\varphi(R)$, $R \bowtie S$. Orthogonally to these definitions, we explore the possibility of annotating each source with quality factors in more than one way. Specifically, we explore the possibility of assigning a single value for the quality of (a) the whole relation, (b) several attributes of the relation (i.e., assuming a *vertical partitioning* of the quality of the relation), and (c) a subset of the rows of the relation (i.e., assuming a *horizontal partitioning* of the quality of the relation). In the sequel, when we refer generically to any of the aforementioned options (i.e., relations, attributes, or subsets of relations) we will call them *relation elements.*

In Fig. 7.19 we present a small motivating example to explain these assumptions. Suppose a multinational manufacturing company producing and selling parts. Suppose also, the existence of different databases in the manufacturing and inventory departments, as well as in the European and American headquarters. In the DSA, we wish to collect all the information about parts and orders. At higher levels of granularity, we want to answer user queries on the combination of these relations. Suppose the following schemata:

ORDER, $ORDER_{A,E}$	(ORD_DATE, CUST_ID, PART_ID, QTY, AGREED_UNIT_PRICE, DELIVERY_DATE)
$PART_I$	(PART_ID, INV_DATE, AVAIL_QTY)
$PART_M$	(PART_ID, MAN_DATE, PROD_QTY, UNIT_COST)
PART	(PART_ID, MAN_DATE, PROD_QTY, UNIT_COST, INV_DATE, AVAIL_QTY)

To compose the information on PARTS in the DSA, we need to join the inventory and manufacturing tables on the primary key PART_ID. In this case, the quality of columns PROD_QTY and AVAIL_QTY is different (i.e., we have a horizontal fragmentation of quality). In the case of ORDERS, where the DSA table is the union of the source tables, the tuples coming from the European database are of a different quality from the tuples coming from the American one; thus we have a case of horizontal fragmentation of quality.

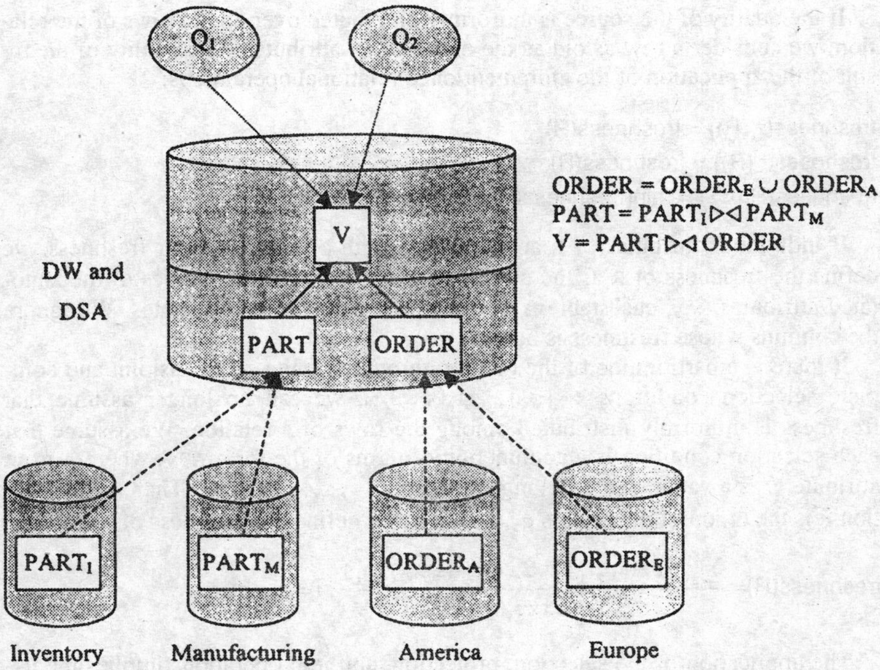

Fig. 7.19. Motivating example

We now provide definitions and metrics for the above quality factors. For cases (b) and (c) we also provide formulas that compute the value of the quality factors for the whole relation, based on its elements. For all three cases, we explore the behavior of the quality of a relation, when a relational operator is applied to it.

Freshness. We measure freshness as the age of the information stored in the data warehouse. For each relation element `r` we assume a maximum time delay for its refreshment: *max delay*(`r`) is the longest period that `r` is left without being updated, and we still accept it as valid for querying purposes. Then, freshness can be defined as:

$$\text{freshness(r)=max}(0, \frac{\text{max delay - time elapsed from last update}}{\text{max delay}})$$

When a relation element is out of date, then its freshness is considered to be zero, whereas it is considered to be 1, exactly after the last update.

In the sequel, we will explore the behavior of freshness, under the application of the select, project, and join operations over a set of underlying relation elements. We assume that **A** is a set of attributes and φ is a selection condition defined as a conjunction of terms of the form $A\theta v$, where A is an attribute, v is a value and θ belongs to the set $\{=, \leq, \geq, \neq, <, >\}$.

If the quality of the source is uniformly distributed over all the rows of the relation, we consider a row as old as the oldest of its attributes. The quality of the result of the application of the aforementioned relational operators is:

freshness($\bullet_A$(R)) = freshness(R)
freshness($\bullet_c$(R)) = freshness(R)
freshness(R$\bowtie$S) = min(freshness(R),freshness(S))

If individual attributes of R are annotated with a value for their freshness, we define the freshness of R as the minimum of all the freshness values of the annotated attributes, i.e., each table is as old as the oldest of its attributes. We ignore the columns whose freshness is not set.

If there is a partitioning of the relation through a finite set of disjoint and complete selection conditions $\Phi=\{\varphi_i\}$, $i=1,..,m$, we can no longer assume that freshness is uniformly distributed among the rows of a relation. We assume that each selection condition is a conjunction of terms of the form $A\theta v$, where A is an attribute, v is a value, and θ belongs to the set $\{=,\leq,\geq,\neq,<,>\}$. Then, if the relation R is the union of the groups g_i, $i=1..m$, we define the freshness of R as:

$$\text{freshness}(R)=\frac{\sum_i \text{size}(g_i) * \text{freshness}(g_i)}{\sum_i \text{size}(g_i)}, i = 1,...,m$$

The application of the selection, projection, and join operations implies the usage of the freshness formulas for the computation of the freshness of each group of the relation R. When combining operations, projection and selection operations that exclude certain groups also influence overall freshness of the results. In the case of join, the freshness of a groupwise join subresult is again the minimum of the freshness of the participating groups; existing results on automatic reasoning for possible incompatibilities between groups can be exploited in this case too [OzVa99], in order to eliminate meaningless groups in the result.

Consistency. We measure consistency as the percentage of the data of a relation that abide by the integrity rules of the database. As one of the reasons for the existence of the DSA is data cleaning, integrity rules are not always enforced by declaration but rather checked by scanning the contents of the imported data.

Again, we shall consider the three options that we explored in the case of freshness. Suppose a vertical partitioning for the quality of a relation $R(A_1,A_2,...,A_n)$, i.e., some attributes of R are annotated with a value for their consistency. This means that we count the distinct number of values of each column and derive the percentage of acceptable ones as the consistency of the column. One could assign different measures to compute the overall consistency of a table from the individual values of the consistency of the columns, e.g., the minimum, the average, or the product of the individual consistencies (in which case one assumes the independence of the values of the columns of the relations). To avoid making any of these assumptions we define the consistency of the relation as the weighted sum of the individual values for the consistency of the columns of the relation. The weights are arbitrarily set by the designer. Average and minimum are trivially covered by this modeling:

$$\text{consistency(R)} = \sum_1^n w_i * \text{consistency}(A_i) = 1, \sum_1^n w_i = 1$$

If we assign different values for consistency to subsets of the rows of a relation, we can estimate the overall consistency of the table as a weighted sum of the individual values, with the size of each group playing the role of weight. Supposing that we have m different groups of rows, the relation R is the union of the groups g_i, i=1..m and the overall consistency of the relations R, is given by:

$$\text{consistency(R)} = \frac{\sum_i \text{size}(g_i) * \text{consistency}(g_i)}{\sum_i \text{size}(g_i)}, i = 1,..,m$$

Next, we proceed to investigate the effects of the application of the selection, projection, and join operations over a relation with respect to its consistency.

If consistency is defined over the whole relation, we assume that the quality of the source is uniformly distributed over all the rows of the relation. The quality of the result of the application of the aforementioned relational operators is then:

consistency($\bullet_A$(R)) = consistency(R)
consistency($\bullet_.$(R)) = consistency(R)
consistency($R \bowtie S$) = consistency(R)*consistency(S)

If consistency is defined over the attributes of a relation, we have a value consistency(A_i) for each attribute. We denote the result of the application of an operator as consistency(A_i'). Then, projection and selection do not change consistency values. For joins, we have

If $R \bowtie_{A=B} S$,
(a) consistency(A_i')=consistency(A_i') for all $A_i \neq$A,B and
(b) consistency(A_i')=consistency(C) for A_i=A',B' and C$\in${A,B} s.t.
distinct(C)=min(distinct(A),distinct(B))

Here, we assume that there is no correlation for the columns that are not involved in the join; thus their consistency is not affected. As far as the columns involved in the join are concerned, the number of distinct values of the result is determined by the column with the fewer distinct values. Suppose B is this column (i.e., C=B for formula 23). If we denote the selectivity of the join as j, the number of distinct values of the result is j*distinct(B) and the number of consistent ones is j*distinct(B)*consistency(B).

If consistency is defined over groups of a relation, we can apply the analogous arguments as for freshness, i.e., there is no serious impact of projection and selection on group consistency, and a join result is only as consistent if both components were consistent. Thus we have the following:

consistency(r_i')=consistency($\bullet_A$(r_i))=consistency(r_i)
consistency(r_i')=consistency($\bullet_.$(r_i))=consistency(r_i)
consistency($r_i \bowtie s_i$)=consistency(r_i)*consistency(s_i))

Accuracy and completeness. We measure accuracy as the percentage of the data of a relation that capture the actual, real-world values of the entities they represent. Measuring accuracy is not a simple issue: sampling techniques [OlRo90] seem to be the best way to perform some kind of measurement. However, the formulas regarding accuracy are identical to the ones regarding consistency.

The same holds for completeness, defined as the fraction of stored values over the number of expected values. In real systems, administrators normally know the size of data they are expecting to receive/produce from a database. Moreover, one can also automate the calculation of this quality factor by counting NULL values in the proper columns.

Summarizing, the DSA example illustrates how relatively simple quantitative models can be developed to optimize the physical design of a data warehouse. For general data warehouse design, this solution must be (a) extended to the handling of views with aggregate functions, and (b) embedded in the qualitative, logic-based design techniques developed in Sect. 3.6 of this book. The current state of research on these rather difficult issues will be covered in the next chapter.

8 Quality-Driven Data Warehouse Design

This final chapter links the metadata framework presented in Chap. 7 with the algorithms presented in earlier chapters to describe a systematic solution to incorporate quality aspects into design problems encountered in data warehousing. This solution, developed in the DWQ project, will be illustrated through an example. Then, one missing building block not yet covered in earlier chapters – the optimization of view materialization – will be discussed with an outlook on current challenges and extensions to data warehousing.

8.1 Interactions between Quality Factors and DW Tasks

As already mentioned in Chaps. 1 and 2, data warehouse (DW) design is one of the most crucial processes in the lifecycle of a data warehouse. In Chaps. 3–6, we have already addressed a number of issues. In Chap. 3, we have addressed the problem of semantic reconciliation of the sources, as well as with the production of an enterprise model for the data warehouse. The solutions suggested there are complemented by Chap. 5, which discussed the semantic modeling of multidimensional databases and the most common logical models for data warehouses (star and snowflake schemata). The role of the operational data store and its specific requirements were investigated in the chapter of update propagation. Finally, in the chapter of query optimization, we discussed the role of auxiliary structures, such as indexing and physical design choices for the quicker processing of user requests from a data warehouse.

All these solutions relate to different data warehouse components, involve interacting logic-based and quantitative models, and contribute in various ways to the different quality goals presented in Chap. 7. To set the stage for our methodology proposal, we briefly summarize the main interactions.

The interpretability of the data and the processes of the data warehouse are heavily dependent on the design process (the level of the description of the data and the processes of the warehouse) and the expressive power of the models and the languages that are used. Both the data and the systems architecture (i.e., where each piece of information resides and what the architecture of the system is) are part of the interpretability dimension. The integration process is related to the interpretability dimension, by trying to produce minimal schemata. Furthermore, processes like query optimization (possibly using semantic information about the kind of the queried data – e.g., temporal, aggregate, etc.), and multidimensional aggregation (e.g., containment of views, which can guide the choice of the appropriate relations to answer a query) are dependent on the interpretability of the data and the processes of the warehouse.

The accessibility dimension of quality is dependent on the kind of data sources and the design of the data and the processes of the warehouse. The kind of views stored in the warehouse, the update policy, and the querying processes are all influencing the accessibility of the information. Query optimization is related to the accessibility dimension, since the sooner the queries are answered, the higher the transaction availability is.

The extraction of data from the sources also influences (actually determines) the transaction availability of the data warehouse. Consequently, one of the primary goals of the update propagation policy should be to achieve high availability of the data warehouse (and the sources).

The update policies, the evolution of the warehouse (amount of purged information), and the kind of data sources all influence the timeliness and, consequently, the usefulness of data. Furthermore, the timeliness dimension influences the data warehouse design and the querying of the information stored in the warehouse (e.g., the query optimization could possibly take advantage of possible temporal relationships in the data warehouse).

The believability of the data in the warehouse is obviously influenced from the believability of the data in the sources. Furthermore, the level of the desired believability influences the design of the views and processes of the warehouse. Consequently, the source integration should take into account the believability of the data, whereas the data warehouse design process should also take into account the believability of the processes. The validation of all the processes of the data warehouse is another issue, related to every task in the data warehouse environment and especially to the design process.

Redundant information in the warehouse can be used from the aggregation, customization, and query optimization processes in order to obtain information faster. Also, replication issues are related to these tasks.

Finally, several factors of data warehouse design are influenced by quality aspects. For instance, the required storage space can be influenced by the amount and volume of the quality indicators needed (time, believability indicators, etc.). Furthermore, problems like the improvement of query optimization through the use of quality indicators (e.g., ameliorate caching), the modeling of incomplete information of the data sources in the data warehouse, the reduction of negative effects schema evolution has on data quality, and the extension of data warehouse models and languages, so as to make good use of quality information have to be dealt with.

8.2 The DWQ Data Warehouse Design Methodology

To achieve a data warehouse design that addresses the full variety of quality issues according to all three perspectives presented in Chap. 7. Individual techniques for particular design issues must be embedded in a seamless methodology that manages the interplay of the different tasks and perspectives. The metadata repository described in Sect. 7.2 serves as the information exchange mechanism to reach these objectives at the data integration level. This section describes how the algorithmic approaches described in other chapters of this book can build on each other. The methodology has been developed in the DWQ project, using a collection of prototypical tools for support [JQC*00].

The DWQ design method consists of the six steps shown in Fig. 8.1. In this figure, the objects of the classical DW architecture have been overlaid with their conceptual correspondences, such as source models (e.g., S1), the enterprise model, and client models (e.g., C1). These conceptual objects are, in our approach, externally represented as extended entity-relationship models, and internally modeled using description logic formalisms from artificial intelligence to allow for subsumption reasoning. The two shaded texts describe related support at the operational level: aggregate query optimization and view refreshment. The whole process is administered through the metadata repository models presented in Sect. 7.2. In the following subsections, we illustrate the main steps by simplified examples taken from a case study for a European telecom organization [TrLN99].

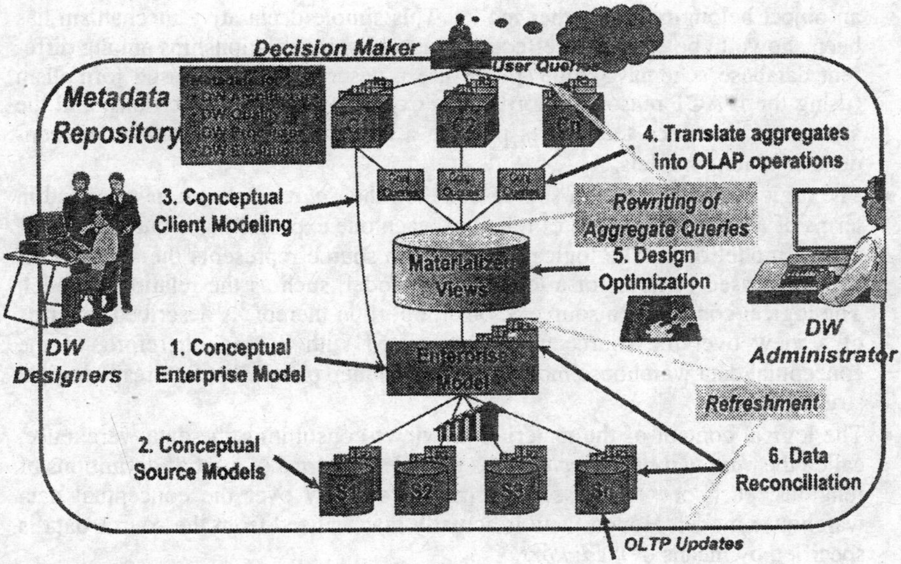

Fig. 8.1. DWQ data warehouse development process

8.2.1 Source Integration

The source integration **steps 1 and 2** in Fig. 8.1 [CDL*01] are not performed as a big 'enterprise modeling' project but are usually intertwined and incremental: whenever a new portion of a source is taken into account, the new information is linked to an evolving "enterprise model." Recall from Chap. 3 that our approach builds on the following submodels:

- The *enterprise model* is a conceptual representation of the global concepts and relationships that are of interest to the data warehouse application. It provides a consolidated view of the concepts and relationships that are important to the enterprise and have so far been analyzed. Such a view is subject to change as the analysis of the information sources proceeds. The description logic formalism we use is general enough to express the usual database models, such as the entity-relationship model and the relational model. Inference techniques associated with the formalism allow for carrying out several reasoning services on the representation.

- For a given information source S, the *source model* of S is a conceptual representation of the data residing in S. Again, the approach does not require a source to be fully conceptualized. Source models are expressed by means of the same formalism used for the enterprise model.
- Integration does not simply mean producing the enterprise model but rather being able to establish the correct relationships between both the source models and the enterprise model and among the various source models. We formalize the notion of interdependency by means of *intermodel assertions* [CaLe93]. An intermodel assertion states that one object (i.e., class, entity, or relation) belonging to a certain model (either enterprise or source model) is always a subset of an object belonging to another model. This simple declarative mechanism has been shown to be extremely effective in establishing relationships among different database schemas [Hull97]. We use a description logic-based formalism (using the iFACT reasoner [Horr99]) to express intermodel assertions, and the associated inference techniques provide a means to reason about interdependencies among models.
- The logical content of each source S, called the *Source Schema*, is provided in terms of a set of definitions of relations, each one expressed as a query over the source model of S. The logical content of a source represents the structure of data expressed in terms of a logical data model, such as the relational model. The logical content of a source S, or of a portion thereof, is described in terms of a view over the source model associated with S (and, therefore, of the conceptual data warehouse model). Wrappers map physical structures to logical structures.
- The logical content of the materialized views constituting the data warehouse, called the *data warehouse schema*, is provided in terms of a set of definitions of relations, each one expressed in terms of a query over the conceptual data warehouse model. How a view is actually materialized from the source data is specified by means of *mediators*.

The following tasks work on this structure within steps 1 and 2 of Fig. 8.1:

- *Enterprise and source model construction*. The source model corresponding to the new source is produced, if not available. Analogously, the conceptual model of the enterprise is produced, if not available.
- *Source model integration*. The source model is integrated into the conceptual data warehouse model. This can lead to changes to both the source models and to the enterprise model. Moreover, intermodel assertions between the enterprise model and the source models and between the new source and the existing sources are added. The designer can specify such intermodel assertions graphically and can invoke various automated analyses.
- *Source and data warehouse schema specification*. The source schema corresponding to a new portion of the source is produced. On this basis, an analysis is carried out if the data warehouse schema should be restructured and/or modified.

In all of these tasks, the metadata repository stores the values of the quality factors involved in source and data integration and helps analyze the quality of the design choices. The quality factors of the conceptual data warehouse model and the various schemas are evaluated and a restructuring of the models and the schemas is accomplished to match the required criteria. Figure 8.2 shows an example where a new correspondence relationship between the (upper) enterprise model (upper entity-relationship (ER) model) and a particular source model (lower ER model) is entered via a simple description logic expression (popup window in the middle). The system will, in this case, return a graphical error message identifying an inconsistency with previous definitions.

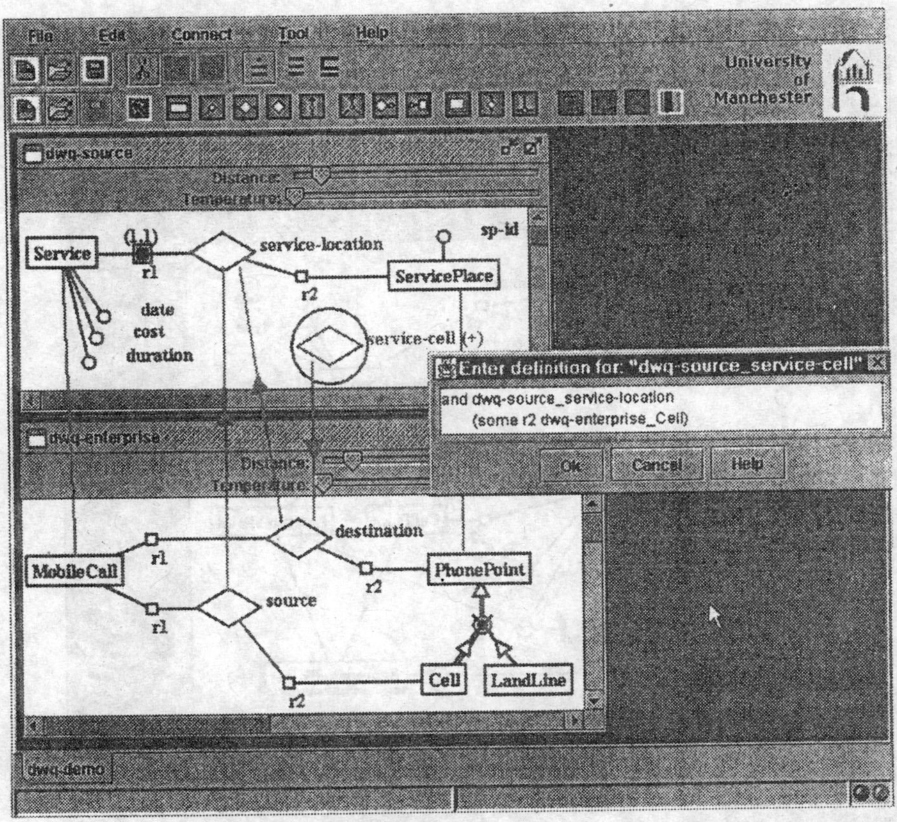

Fig. 8.2. Specification and DL-based quality control of source integration relationships

8.2.2 Multidimensional Aggregation and OLAP Query Generation

The next two steps consider the client. Again, these steps can be performed incrementally, as new client interests are recognized.

Step 3: Conceptual Client Modeling. The conceptual modeling language underlying the enterprise and source models and the corresponding modeling and reasoning tools have been extended to the case where concepts are organized into aggregates along multiple dimensions with multihierarchy structure [FrSa99]. Data warehouse designers can thus define multidimensional and hierarchical views over the enterprise conceptual model in order to express the interests of certain DW client profiles, without losing the advantages of consistency and completeness checking as well as semantic optimization provided by the conceptual modeling approach. Examples for such multihierarchies include time (modeled in days-weeks or in days-months-years). Figure 8.3 gives an impression of this extension in which a new aggregate relationship (Aggregate-MobileCalls2) is inferred from existing ones and can then be renamed adequately.

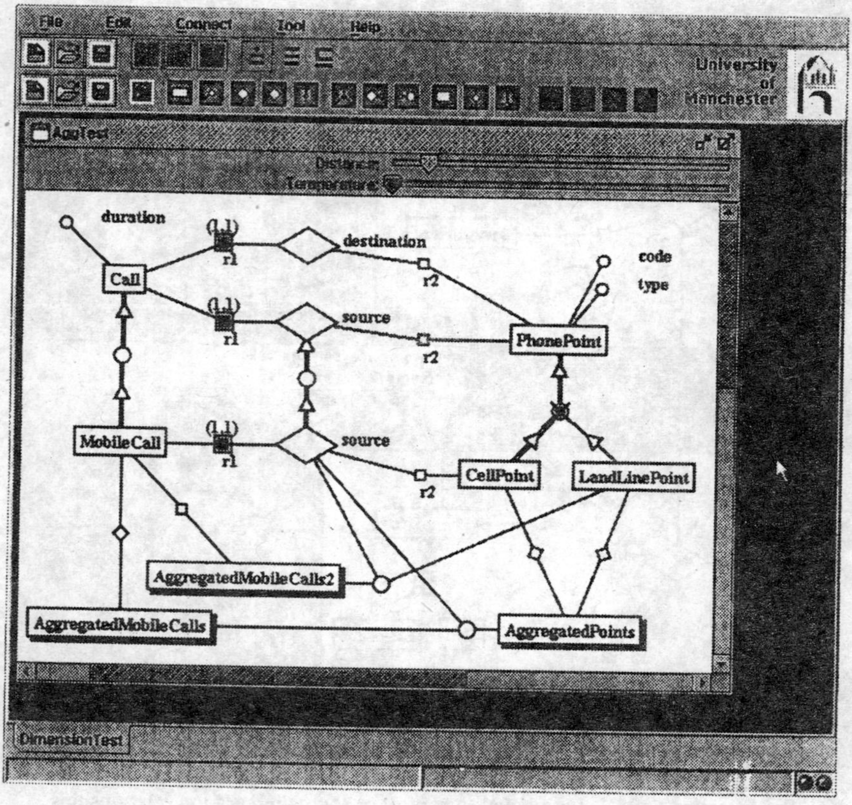

Fig. 8.3. Extended ER model and DL reasoning facility enhanced with aggregates

Step 4: Translate aggregates into OLAP queries. The thus defined "conceptual data cubes" can either be implemented directly by multidimensional (MOLAP) data models or supported by a mapping to relational database systems (ROLAP). A star schema is often used in OLAP systems with very large sparse data cubes because a mapping to just one data cube would be extremely inefficient and therefore the nonempty pieces of the cube are parceled out into many smaller subcubes. Faithful representation of client views requires a careful design of an OLAP relational algebra, together with the corresponding rewritings to underlying star schemas [Vass98].

8.2.3 Design Optimization and Data Reconciliation

Step 5: Quantitative design optimization. In the logical perspective, our conceptual approach to source integration has created a schema of relations implementing the enterprise model. This schema could be implemented directly as an operational data store (ODS). In a data warehouse with many updates and unpredictable queries, this would be the appropriate view materialization.

On the other extreme, the mapping of multidimensional aggregates to ROLAP queries creates a set of view definitions (queries). The materialization of these queries would be the optimal solution (storage space permitting) in a query-only data warehouse with hardly any updates.

Typical data warehouses have less extreme usage patterns and therefore require a compromise between the two view materialization strategies. Solutions for this problem are described in Sects. 7.5 and 8.3.

Step 6: Data reconciliation. The physical-level optimization is fully integrated with the conceptual modeling approaches because it works on their outcomes. Conversely, the resulting optimal design is now implemented by data integration and reconciliation algorithms, semiautomatically derived from the conceptual specifications. The views to be materialized are initially defined over the ODS relations. There can be several qualitatively different and possibly conflicting ways to actually materialize these ODS relations from the existing sources. These ways are generated by further rewritings derived from the source integration definitions.

As discussed in Sect. 3.2, the problem of data reconciliation arises when data passes from the application-oriented environment to the data warehouse. During the transfer of data, possible inconsistencies and redundancies are resolved, so that the warehouse is able to provide an integrated and reconciled view of data of the organization. In [CDL*01], data reconciliation is based on (1) specifying through interschema assertions how the relations in the data warehouse schema are linked to the relations in the source schemas, and (2) designing suitable mediators for every relation in the data warehouse schema.

In step 1, interschema correspondences are used to declaratively specify the correspondences between data in different schemas (either source schemas or data warehouse schema). Interschema correspondences are defined in terms of relational tables, similarly to the case of the relations describing the sources at the logical level. We distinguish three types of correspondences, namely conversion, matching, and reconciliation correspondences. By virtue of such correspondences, the designer can specify different forms of data conflicts holding between the sources and can anticipate methods for solving such conflicts when loading the data warehouse.

In step 2, the methodology aims at producing, for every relation in the data warehouse schema, a specification of the corresponding mediator, which determines how the tuples of such a relation should be constructed from a suitable set of tuples extracted from the relations stored in the sources.

In effect, we have here a semiautomatic generation and embedding of mediator agents for data extraction, transformation, cleaning, and reconciliation. Figure 8.4 shows how the prototypical tools developed in the DWQ project solve this problem by query adornments in a deductive database style which maps easily to standard relational query languages such as SQL. It may be noteworthy to see that the solution involves the use of elementary operations, such as convert or match,

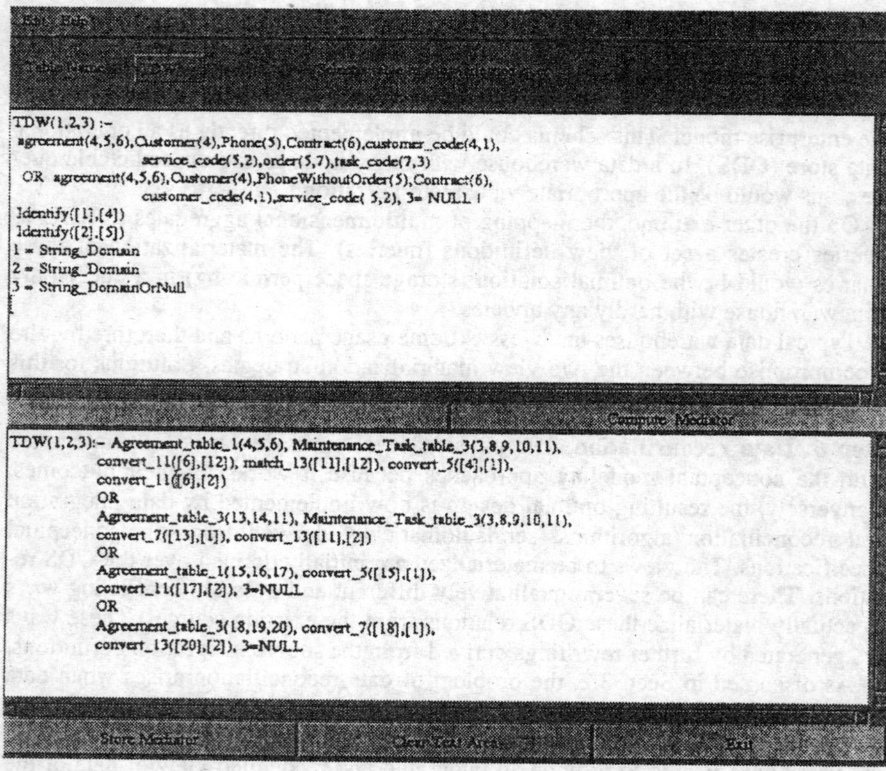

Fig. 8.4. Data reconciliation using mediator frameworks with adorned logic queries

which have also been postulated as basic ingredients of an algebra for the generic management of metadata models [BeRa00].

8.2.4 Operational Support

The data warehouse run-time environment, which operates on the structures and mechanisms designed as described in the previous subsections, needs enhancements both at the logical and physical levels.

At the logical level, the system needs to take advantage of the design optimization in step 4 which generated a specific set of materialized views to reduce query costs. However, the user should not have to know which views have been materialized. Indeed, ideally the user should be able to formulate the queries directly on graphical or form-based representations of the conceptual client views designed in step 3. This creates the need for a further automatic rewriting – from the OLAP mapping of the client views (prepared in step 4) to the views materialized in step 5. As discussed in Sect. 6.2, this rewriting problem becomes quite tricky where aggregates are involved, but good solutions (some of them heuristics) are known for a number of special cases. In [CoNS99], the solution underlying our approach is described. An example is given in Fig. 8.5.

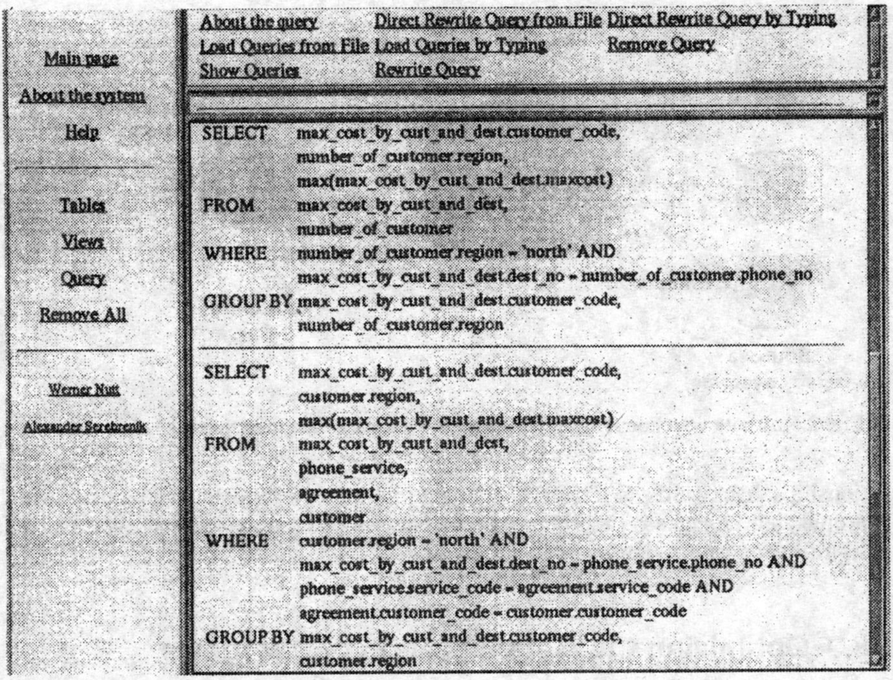

Fig. 8.5. Query rewriting using aggregate views

As discussed in Chap. 4, source data are continuously changing – one of the main reasons why triggers and agent technologies have been proposed for monitoring such changes through active wrappers and reconciling them incrementally into the data warehouse. Issues here do not just involve the logical perspective but also physical aspects of performance and reliability. We shall not discuss these issues here but direct the reader to [NiJa00] for a survey of modeling techniques for replicated and distributed databases) and to Sect. 4.5 for the workflow approach taken in DWQ.

A case study [JQB*99] of how the logical and physical aspects of refreshment can be integrated for innovative solutions in practice is shown in Fig. 8.6. The enterprise in question regularly extracts heterogeneous transactional source data into a (relational) data warehouse of information relevant for its sales force. They have implemented some of our view maintenance mechanisms [StJa96] and combined them with the loading and replication mechanisms offered by the groupware system Lotus Notes at the physical level to create mobile agents which can be carried along on the laptops of the sales force and resynchronized from time to time with the status of the data warehouse.

However, the declarative approach pursued by us for the mapping among sources, enterprise, and client models led to another surprising effect. Sales people noted that the quality of the data produced by the data warehouse was much better than that of the operational data sources due to the reconciliation mechanisms offered. The same sales people also produced transactions for these sources – with much less quality control. As a consequence, we used the same model inter-rela-

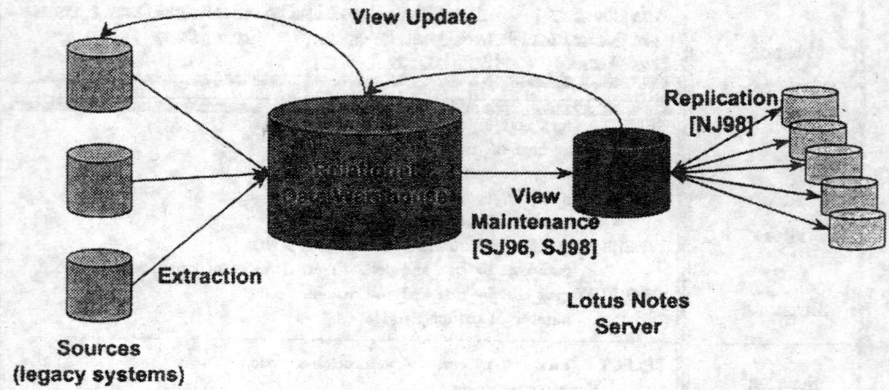

Fig. 8.6. A data warehouse with forward and backward data refreshment

tionships backwards to create a semiautomatic view update mechanism which propagated changes on the client views back into the transactional systems, thus significantly improving source data quality.

8.3 Optimizing the Materialization of DW Views

While most of the ingredients of the just presented DWQ approach were already discussed in earlier chapters, we still must fill one gap related to the quantitative optimization step 4 of our methodology beyond the approach presented in Sect. 7.5. Are the traditional schemata sufficient for the efficient processing and update of the information residing in a DW in the presence of aggregations?

The basic assumption behind most of the work below is that a ROLAP model exists, where a fact table with all the detailed information provides the basis for further aggregations. It must be critically noted that this assumption is not true for the majority of OLAP systems currently deployed, but, unfortunately, research has presented few solutions to the corresponding problem within the more frequently used MOLAP domain. Nevertheless, it can be hoped that some of the results presented below can in fact be transferred to MOLAP with minor modifications.

Kimball [Kimb96] suggests storing, if possible, aggregates for all combinations of attributes. This, however, may lead to undesirable effects. First, the number of views to be stored is the product of the numbers of attributes for each dimension, so a large number of views may have to be stored. Second, the views may be of size not much smaller than the original data cube. This can happen if the data cube is sparse, i.e., if there are only facts for few combinations of dimension objects. In such a case, the classes in the partition of facts defined by a choice of attributes may have only a few elements, and there may be almost as many nonempty classes of facts as there are facts. Kimball proposes, therefore, to store only those views where aggregate classes have at least 10–20 elements on average.

Harinarayan et al. [HaRU96] give a refined analysis of which precomputed views give the best benefit when answering aggregate queries and which views should be materialized if there are space constraints. They assume that a data cube

C is given, over dimensions $D_1,...,D_n$, each of which comes with a set of attributes. The attributes of a dimension form a hierarchy. We write $a \geq b$ if attribute a has a finer granularity than b or the same. A *derived cube* $C_{a1,...,an}$ is determined by choosing exactly one attribute a_i from each dimension D_i and performing aggregation with respect to these attributes. We write $C_{a1,...,an} \geq C_{b1,...,bn}$ if the cube $C_{a1,...,an}$ can be computed from the cube $C_{b1,...,bn}$. This is the case if and only if $a_i \geq b_i$ for $i = 1,...,n$, since then the classes in the partition with respect to b_i are unions of classes in the partition with respect to a_i. Thus, the derived cubes also form a hierarchy.

As an example, suppose two dimensions Customer and Product are given with attribute hierarchies as in Fig. 8.1. Figure 8.2 illustrates the hierarchy on the derived cubes that is inherited from the one on the attributes.

There are two observations about the cube hierarchy:

- A cube C can be used to compute C' if C is above C' in the cube hierarchy. Hence, the *higher* a cube, the *more useful* it is.
- The *higher* subcubes are likely to consume *more space.*

These requirements create a design conflict if there are space restrictions. In order to find an optimal solution one has to estimate the expected size of a cube and the speedup for the queries that results from its materialization. Harinarayan et al. [HaRU96] show that the problem of selecting, under space constraints, a set of cubes that guarantees a certain speed up is NP-complete. They also show that a greedy algorithm that always materializes the most promising cube next finds a solution that yields at least 63% of the optimal speedup. Their results are extended for the case where different probabilities for view usage exist. An analysis for the time/space tradeoff is also provided.

If space is restricted, the space occupied by indexes should also be considered. In [GHRU97], the problem of simultaneously choosing views and indexes, on the base data and on the views, is investigated. The modeling is not much different from [HaRU96], since a view is defined with respect to its grouping attributes; yet its selection condition is also taken into account. The cost model is extended from the linear cost model to a more complex one, but the two resulting greedy algorithms do not guarantee the same results as with the algorithm of [HaRU96].

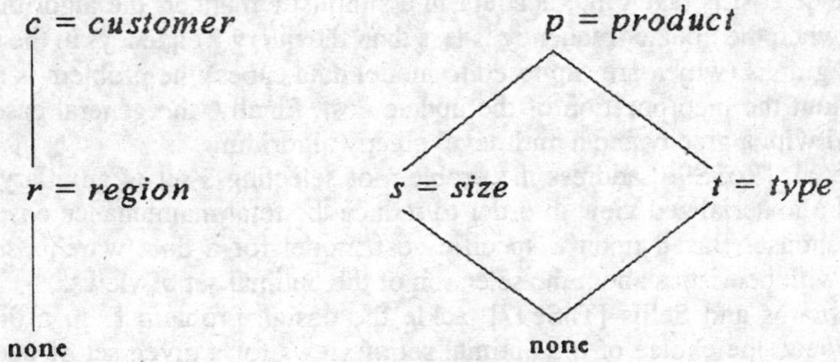

Fig. 8.7. The customer and the product hierarchy [HaRU96]

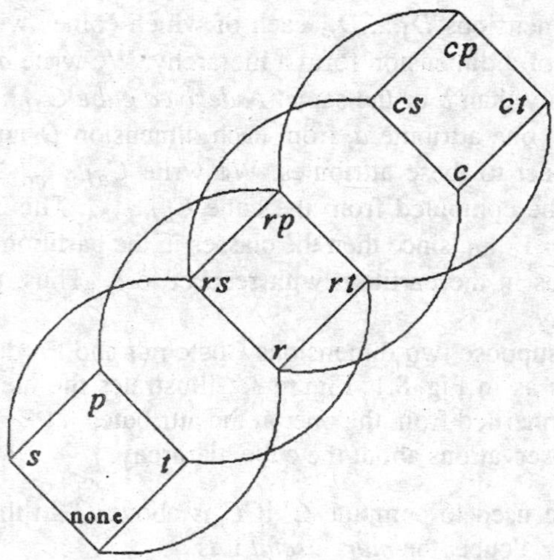

Fig. 8.8. Combining the customer and the product hierarchy [HaRU96]

The follow-up paper in this line of research, [Gupt97], takes into account both the refreshment and the querying cost. The modeling is also different: views are represented as AND-OR graphs. A view defined as an *AND view graph* involving other views or base relations requires the participation of all of its "subviews" in order to be correctly computed. A view defined as an *OR view graph* has multiple plans for its evaluation – each one can be an AND view graph. The problem again involves the case when we have prior knowledge of the user queries, and we want to come up with the optimal set of views in terms of total operational cost. The first step to take is the construction of the input graph for the problem: for each query, all the alternative evaluation paths are explored and merged in a single graph. The problem is proved to be NP-hard and special cases are considered. In the case of AND view graphs (where there is a unique way for the computation of a view) several variations of a greedy algorithm are proposed, involving the selection of indexes along with the respective views and the extension of the original greedy algorithm with a *greedy interchange* algorithm (improving the result of the greedy algorithm by replacing a view with another, not originally selected). The maintenance cost is taken into account in a simplistic manner: the algorithms apply only when the update frequency is less than the query frequency. In the case of OR view graphs (which are supposed to model data cubes) the problem is considered without the incorporation of the update cost. Finally, the general case is investigated with a greedy and a multilevel greedy algorithm.

Ross et al. [RoSS96] address the problem of selecting a set of auxiliary views on top of a materialized view in order to reduce the total maintenance cost of the data warehouse. Based upon a specific cost model for a data warehouse, they come up with heuristics about the selection of the optimal set of views.

Theodoratos and Sellis [ThSe97] tackle the design problem from a different point of view: the choice of the optimal set of views for a given set of queries is modeled as a state space problem. Only select-join queries and views are considered. Views are represented by multiquery graphs. A state is also a multiquery graph for the views to be materialized in the data warehouse and a complete rewriting of the queries over these views. An exhaustive algorithm making transitions is proposed: a transition from one state to another transforms the graph and rewrites the queries over the new graph. The selection of a transition is based on a cost model that takes into account both the update and the querying cost. The cost model is generic, so that it can be customized to more specific models, without affecting the algorithm. Finally, two heuristics are provided for the pruning of the search space of the algorithm.

Yang et al. [YaKL97] customize previous work on multiple query optimization for the problem of optimal data warehouse design. For a given set of queries, expressed over a set of base relations, a global query access plan is produced, in which the local access plans for individual queries are merged, based on the shared operations on common data sets. A generic cost model is also provided, which takes into account both the query frequencies and the base relations update frequencies. There are three algorithms provided; the first and the second use techniques from single and multiple query optimization, extended with heuristics for the pruning of the search space. The second algorithm is actually an extension of the first, where instead of the global multiple query graph, the queries are primarily optimized locally. The third algorithm uses a mapping of the problem to a 0–1 integer programming problem so that the solution is guaranteed to be optimal.

In [BaPT97], the relations participating in a DW schema are distinguished as *dimension* and *fact* tables. A multidimensional database constitutes a set of dimension tables and a fact table related to them through foreign keys. Attribute hierarchies are expressed as functional dependencies on the attributes of the dimensional tables. A *cube lattice* is the set of all the possible grouping queries that can be defined on the foreign keys of a fact table (much like the lattice introduced in [HaRU96]). A query q is the *ancestor* of a query q' if all its grouping attributes are at a higher level of aggregation than the respective attributes of q'. Based on this relationship, an *MD-lattice* of attributes is built, modeling the combination of all the attributes participating in a multidimensional database. Again, given a set of queries, the problem is the minimization of the total query and update cost, where all the queries are answered. An algorithm is proposed for the solution of the problem, based on a hypothesis of monotonicity: a detailed view must have a bigger operational cost than a more detailed one for the algorithm to produce correct results. The algorithm chooses the candidate views based on their usability with respect to the queries participating to the problem. Then, the set of the views which are to be materialized is chosen by taking into account their least upper bound in the MD-lattice. Heuristics for the pruning of the MD-lattice are also suggested.

In [SCJL98], a method for representing a data cube in terms of different kinds of view elements is proposed. View elements can either be *aggregated, intermediate*, or *residual*. Aggregation operators such as *partial sum* and *total sum* are proposed. These aggregation operators fulfill properties such as perfect reconstruction (of views from their subviews), nonexpansiveness (of the volume of the data cube), distributivity, and separability. The problem is formulated as a view element graph. An algorithm is provided for the selection of the optimal set of views with respect to the minimization of the query processing cost. A greedy algorithm complements the previous solutions with a set of auxiliary views in order to minimize the processing and storage cost.

8.4 Summary and Outlook

Concluding, it is useful to summarize the state of the art for quality in data warehouse environments. We have explained why the need for quality in data warehouses is eminent. Despite the enormous variety of individual techniques for source integration, data refreshment, multidimensional modeling, and querying discussed throughout the book, theory and practice offer few tools to investigate and understand explicitly the quality of data. Most of these tools are customized to specific applications (e.g., customer name and address manipulation) and are rather poorly integrated.

In this final chapter, we therefore adapted some well-known approaches from industrial quality management, data quality modeling, and software quality assessment to come up with an extension of metadata management in data warehousing which will enable DW designers and administrators to answer question such as:

- What is the methodology to model quality for a data warehouse?
- What are the dimensions of data and process quality for a data warehouse?
- How do we measure quality in a data warehouse environment?
- How do we relate quality to every aspect of a data warehouse?

The DWQ framework and metadata model provide an overall conceptual umbrella under which these issues can be discussed and interrelated in a meaningful manner. We showed how the DWQ methodology and prototypical tool set makes use of these metadata for a seamless approach to quality-driven data warehouse design and operation. Industrial case studies with major portions of the approach seem to support this claim.

Recently, the data warehouse approach has come under attack as being too heavy and centralized in the age of virtual organizations. Throughout this book, it has been pointed out, however, that nothing prevents a physical or even logical distribution of the data as long as their conceptual relationships are reasonably well understood. The conceptual business perspective introduced by the DWQ framework and the conceptual modeling and reasoning techniques associated with it offer a way to combine distributed implementations of data warehousing, such as data marts without a central data warehouse, with a global conceptual understanding of their meaning for the organization.

Indeed, concept-centered techniques for data integration and client delivery are likely to become even more useful when applied in information brokering settings which are conceptually less centralized and technically much more distributed than data warehouses. A very strong example is the current move towards a Semantic Web where standards organizations such as the World Wide Web consortiums are adopting conceptual modeling standards with description logic foundations developed in part by co-authors of this book in order to enable more efficient information integration and search in very large multimedia information networks such as the Internet. While it may be doubtful if such conceptual models can be developed globally and in a domain-independent manner, there is great promise that they might strongly improve the efficiency of domain-specific supply chains or electronic markets, and several research consortia and companies throughout the world are working to make this happen.

Another growing area where conceptualization may significantly improve quality as well as efficiency is data mining. It is often cited that typical data warehousing tasks, such as information collection and data cleaning, account for at least 85% of data mining work. It is therefore not surprising that solutions inspired by the data warehouse idea are diffusing into discovery-intensive fields, such as web clickstream analysis; scientific high-performance computing in physics, satellite observations, and the like; and in engineering processes for experience reuse. These applications also feed back into the evolution of data warehouse research and practice itself (e.g., data mining can be used as a means for automatically enriching source and enterprise models, thus, enabling better data cleaning), such that many further data warehouse innovations can be expected in the next years.

Bibliography

[AAD*96] S. Agarwal, R. Agrawal, P. Deshpande, A. Gupta, J. Naughton, R. Ramakrishnan, S. Sarawagi: On the computation of multidimensional aggregates. In: [VLDB96], pp. 501–521.

[AbGr95] Aberdeen Group: Data warehouse query tools: Evolving to relational OLAP. White Paper at the Aberdeen Group, 1995.

[Abit97] S. Abiteboul: Querying semi-structured data. In: [ICDT97], pp. 1–18.

[ACHK93] Y. Arens, C. Y. Chee, C. Hsu, C. A. Knoblock: Retrieving and integrating data from multiple information sources. Journal of Intelligent and Cooperative Information Systems, **2**(2), pp. 127–158, 1993.

[AdCo97] AdCom Services: Data Warehouse-Project Proposal. http://pulsar.adcom.uci.edu/dwh/dwpro4/prohome.html, 1997.

[AdGW97] B. Adelberg, H. Garcia-Molina, J. Widom: The STRIP rule system for efficiently maintaining derived data. In: [SIGM97], pp. 147–158.

[AdKG96] B. Adelberg, B. Kao, H. Garcia-Molina: Database support for efficiently maintaining derived data. In: [EDBT96], pp. 223–240.

[AdVe98] C. Adamson, M. Venerable: Data Warehouse Design Solutions. Wiley, 1998.

[AESY97] D. Agrawal, A. El Abbadi, A. Singh, T. Yurek: Efficient view maintenance at data warehouses. http://www.cs.ucsb.edu/%7ecs172/handouts/readings.html, 1997.

[AFGP96] A. Artale, E. Franconi, N. Guarino, L. Pazzi: Part-whole relations in object-centered systems: An overview. Data and Knowledge Engineering, **20**(3), pp. 347–383, 1996.

[AgAh87] N. Agmon, N. Ahituv: Assessing data reliability in an information system. Journal of Management Information Systems, 4(2), 1987.

[AgGS95] R. Agrawal, A. Gupta, S. Sarawagi: Modeling multidimensional databases. Technical report, IBM Almaden Research Center, San Jose, CA, 1995.

[AhSU79] A.V. Aho, Y. Sagiv, J.D. Ullman: Efficient optimization of a class of relational expressions. ACM Transactions on Database Systems, 4(4), pp. 435–454, 1979.

[Akao90] Y. Akao (ed): Quality Function Deployment. Productivity Press. Cambridge, MA, 1990.

[Albe96] J. Albert: Data integration in the RODIN multidatabase system. In: Proceedings of the 4th International Conference on Cooperative Information Systems (CoopIS), Brussels, Belgium, pp. 48–57. IEEE Computer Society, 1996.

[Arbo96] Arbor Software Corporation: Arbor Essbase. http://www.arborsoft.com/essbase.html, 1996.

[ArFr98] A. Artale, E. Franconi: Temporal description logics. In: L. Vila, P. van Beek, M. Boddy, M. Fisher, D. Gabbay, A. Galton, R. Morris (eds): Handbook of Time and Temporal Reasoning in AI. MIT Press, 1998.

[ArKC96] Y. Arens, C. A. Knoblock, W. Chen: Query reformulation for dynamic information integration. Journal of Intelligent Information Systems, **6**(2/3), pp. 99–130, 1996.

[ASII91] American Supplier Institute, Inc.: Taguchi methods: Introduction to quality engineering. Allen Park, 1991.

[BaBM97] J. Bager, J. Becker, R. Munz: Central warehouse (in German). c't (Magazin für Computer Technik), No. 3, 1997.

[BaCP96] E. Baralis, S. Ceri, S. Paraboschi: Conservative timestamp revisited for materialized view maintenance in a data warehouse. In: [MuGu96], pp. 1–9.

[BaHa91] F. Baader, P. Hanschke: A schema for integrating concrete domains into concept languages. In: Proceedings of the 12th International Joint Conference on Artificial Intelligence (IJCAI), Sydney, pp. 452–457. Morgan Kaufmann, 1991.

[BaHa93] F. Baader, P. Hanschke: Extensions of concept languages for a mechanical engineering application. Proceedings of the 16th German AI-Conference (GWAI), Bonn, Germany, LNCS 671, pp. 132–143. Springer, 1993.

[BaLN86] C. Batini, M. Lenzerini, S.B. Navathe: A comparative analysis of methodologies for database schema integration. ACM Computing Surveys, **18**(4), pp. 323–364, 1986.

[BaPa82] D.P. Ballou, H.L. Pazer: The impact of inspector failability on the inspection policy serial production system. Management Science, **28**(4), 1982.

[BaPa85] D.P. Ballou, H.L. Pazer: Modeling data and process quality multi-input, multi-output information systems. Management Science, **31**(2), 1985.

[BaPa87] D.P. Ballou, H.L. Pazer: Cost/quality tradeoffs for control procedures information systems. Omega: International Journal of Management Science, **15**(6), 1987.

[BaPT97] E. Baralis, S. Paraboschi, E. Teniente: Materialized view selection in a multidimensional database. In: [VLDB97], pp. 156–165.

[BaSa98] F. Baader, U. Sattler: Description logics with concrete domains and aggregation. In: Proceedings of the 13th European Conference on Artificial Intelligence (ECAI), pp. 336–340. Wiley, 1998.

[BaTa89] D.P. Ballou, K.G. Tayi: Methodology for allocating resources for data quality enhancement. Communications of the ACM, **32**(3), pp. 320–329, 1989.

[BBBB95] D. H. Besterfield, C. Besterfield-Michna, G. Besterfield, M. Besterfield-Sacre: Total Quality Management. Prentice Hall, Englewood Cliffs, NJ, 1995.

[BBC*99] P.A. Bernstein, T. Bergstraesser, J. Carlson, S. Pal, P. Sanders, D. Shutt: Microsoft Repository version 2 and the open information model. Information Systems, **24**(2), pp. 71–98, 1999.

[BCC*01] C.J. Bunger, L.S. Colby, R.L. Cole, W.J. McKenna, G. Mulagund, D.G. Wilhite: Aggregate maintenance for data warehousing in Informix Red Brick Vista. Proceedings of the 27th VLDB Conference, Rome, Italy, 2000.

[BeRa00] Bernstein, P.A., E. Rahm: Data warehouse scenarios for model management. In: Proc. Intl. Conf. On Conceptual Modeling (ER 2000), LNCS 1920, pp. 1–15. Springer, 2000.

[Best94] D.H. Besterfield: Quality Control. Prentice Hall, Englewood Cliffs, NJ, 1994.

[BFL*97] M. Bouzeghoub, F. Fabret, F. Llirbat, M. Matulovic, E. Simon: Designing data warehouse refreshment system. DWQ Technical Report, October 1997.

[BFM*98] M. Bouzeghoub, F. Fabret, M. Matulovic, E. Simon: A toolkit approach for developing efficient and customizable active rule systems. DWQ Technical Report, October 1998.

[BJNS94] M. Buchheit, M.A. Jeusfeld, W. Nutt, M. Staudt: Subsumption of queries over object-oriented databases. Information Systems, **19**(1), pp. 33–54, 1994.

[BlIG94] J.L. Blanco, A. Illarramendi, A. Goñi: Building a federated relational database system: An approach using a knowledge-based system. Journal of Intelligent and Cooperative Information Systems, **3**(4), pp. 415–455, 1994.

[BlLT86] J.A. Blakeley, P.A. Larson, F.W. Tompa: Efficiently updating materialized view. In: Proceedings of the ACM SIGMOD International Conference on Management of Data, Washington DC, pp. 61–71. ACM Press, 1986.

[BoCo90] M. Bouzeghoub, I. Comyn-Wattiau: View integration by semantic unification and transformation of data structures. In: Proceedings of the 9th International Conference on the Entity-Relationship Approach (ER), pp. 413–430. North-Holland, 1990.

[BoDS00] V. Borkar, K. Deshmuk, S. Sarawagi. Automatically Extracting Structure from Free Text Addresses. Bulletin of the Technical Committee on Data Engineering, **23**(4), pp. 27–32, 2000.

[Boed87] R.F. Boedecker: Eleven Conditions for Excellence: The IBM Total Quality Improvement Process. American Institute of Management, Quincy, MA, 1987.

[Boeh89] B. Boehm: Software Risk Management. IEEE Computer Society Press, 1989.

[Borg94] A. Borgida: Description logics for querying databases: In: F. Baader, M. Lenzerini, W. Nutt, P. F. Patel-Schneider (eds): Working Notes of the 1994 Description Logics Workshop, Bonn, Germany, 1994.

[BrEM95] L. Briand, K. El Emam, S. Morasca: On the application of the measurement theory in software engineering. Technical report, no. 95-04, International Software Engineering Research Network, available at http://www.iese.fhg.de/ISERN/pub/isern_biblio_tech.html, 1995.

[BrMB94] L. Briand, S Morasca, V. Basili: Property-based software engineering measurement. Technical report, CS-TR-119, Univ. of Maryland, November 1994.

[BSHD98] M. Blaschka, C. Sapia, G. Hofling, B. Dinter: Finding your way through multidimensional data models. In: [DEXA98], pp. 198–203.

[BuDK92] P. Buneman, S. Davidson, A. Kosky: Theoretical aspects of schema merging. Proceedings of the 3rd International Conference on Extending Database Technology (EDBT), Vienna, Austria, LNCS 580, pp. 152–167. Springer, 1992.

[BWPT93] D.P. Ballou, R.Y. Wang, H.L. Pazer, K.G. Tayi: Modeling data manufacturing systems to determine data product quality. Technical Report, no. TDQM-93-09, Total Data Quality Management Research Program, MIT Sloan School of Management, Cambridge, MA, 1993.

[BySc96] J. Byard, D. Schneider: The ins and outs (and everything in between) of data warehousing. Tutorials of ACM SIGMOD International Conference on Management of Data. Montreal, Canada, 1996.

[CaDL95] T. Catarci, G. D'Angiolini, M. Lenzerini: Conceptual language for statistical data modeling. Data and Knowledge Engineering, **17**(2), pp. 93–125, 1995.

[CaLe93] T. Catarci, M. Lenzerini: Representing and using interschema knowledge in cooperative information systems. Journal of Intelligent and Cooperative Information Systems, **2**(4), pp. 375–398, 1993.

[CaLN94] D. Calvanese, M. Lenzerini, D. Nardi: A unified framework for class based representation formalisms. In: J. Doyle, E. Sandewall, P. Torasso, (eds): Proceedings of the 4th International Conference on the Principles of Knowledge Representation and Reasoning (KR), Bonn, Germany, pp. 109–120. Morgan Kaufmann, 1994.

[CaTo97] L. Cabibbo, R. Torlone: Querying multidimensional databases. Proceedings of 6th International Workshop on Database Programming Languages (DBPL-6), Estes Park, CO, LNCS 1369, pp. 319–335. Springer, 1997.

[CaTo98] L. Cabibbo, R. Torlone: A logical approach to multidimensional databases. In: [EDBT98], pp. 183–197.

[CaTo98a] L. Cabibbo, R. Torlone: From a procedural to a visual query language for OLAP. In: [SSDB98], pp. 74–83.

[CDL*97] D. Calvanese, G. De Giacomo, M. Lenzerini, D.Nardi, R. Rosati: Source integration in data warehousing. DWQ Technical Report, October 1997.

[CDL*98] D. Calvanese, G. De Giacomo, M. Lenzerini, D. Nardi, R. Rosati: Information integration: Conceptual modeling and reasoning support. In: Proceedings of the 6th International Conference on Cooperative Information Systems (CoopIS), New York, pp. 280–291. IEEE Computer Society, 1998.

[CDL*98a] D. Calvanese, G. De Giacomo, M. Lenzerini, D. Nardi, R. Rosati: Description logic framework for information integration. In: Proceedings of the 6th International Conference on the Principles of Knowledge Representation and Reasoning (KR), pp. 2–13. Morgan Kaufmann, 1998.

[CDL*98b] D. Calvanese, G. De Giacomo, M. Lenzerini, D. Nardi, R. Rosati: Source integration in data warehousing. In: [DEXA98], pp. 192–197.

[CDL*01] D. Calvanese, G. De Giacomo, M. Lenzerini, D. Nardi, R. Rosati: Data integration in data warehousing. International Journal of Cooperative Information Systems, Spring, 2001.

[CDLV00a] D. Calvanese, G. De Giacomo, M. Lenzerini, M. Y. Vardi: Answering regular path queries using views. In: Proceedings of the 16th IEEE International Conference on Data Engineering (ICDE-00), pp. 389–398IEEE Press. 2000.

[CDLV00b] D. Calvanese, G. De Giacomo, M. Lenzerini, M. Y. Vardi: View-based query processing for regular path queries with inverse. In: Proceedings of the 19th ACM SIGACT SIGMOD SIGART Symposium on Principles of Database Systems (PODS-00), pp. 58–66.ACM Press, 2000.

[CDLV00c] D. Calvanese, G. De Giacomo, M. Lenzerini, M. Y. Vardi: View-based query processing and constraint satisfaction. In: Proceedings of the 15th IEEE Symposium on Logics in Computer Science (LICS-00), pp. 361–371 IEEE Press. 2000.

[CeWi91] S. Ceri, J. Widom: Deriving production rules for incremental view maintenance. Proceedings of the 17th International Conference on Very Large Databases (VLDB), Barcelona, Spain, pp. 577–589. Morgan Kaufmann, 1991.

[CGH*94] S. Chawathe, H. Garcia-Molina, J. Hammer, K. Ireland, Y. Papakonstantinou, J. Ullman, J. Widom: The TSIMMIS project: Integration of heterogeneous information sources. In: Proceedings of IPSJ Conference, Tokyo, Japan, pp. 7–18. 1994.

[CGL*96] L. Colby, T. Griffin, L. Libkin, I.S. Mumick, H. Trickey: Algorithms for deferred view maintenance. In: [SIGM96], pp. 469–480.

[ChDa96] S. Chaudhuri, U. Dayal: Data warehousing and OLAP for decision support. In: Tutorials of 22nd International Conference on Very Large Data Bases (VLDB), Mumbai, India, September 1996.

[ChDa97] S. Chaudhuri, U. Dayal: An overview of data warehousing and OLAP technology. SIGMOD Record, **26**(1), pp. 65–74, 1997.

[ChMe77] A.K. Chandra, P.M. Merlin: Optimal implementation of conjunctive queries in relational databases. In: Proceedings of the 9th Annual ACM Symposium on Theory of Computing (STOC), Boulder, CO, pp. 77–90.ACM Press 1977.

[ChRo94] C.M. Chen, N. Roussopoulos: The implementation and performance evaluation of the ADMS query optimizer. Proceedings of the 4th International Conference on Extending Database Technology (EDBT), Cambridge, UK, LNCS 779, pp. 323–336. Springer, March 1994.

[ChRo96] D. Chatziantoniou, K. Ross: Querying multiple features in relational databases. In: [VLDB96], pp. 295–306.

[ChSh94] S. Chaudhuri, K. Shim: Including group-by in query optimization. In: Proceedings of the 20th International Conference on Very Large Data Bases (VLDB), Santiago de Chile, Chile, pp. 354–366. Morgan Kaufmann, August 1994.

[ChVa93] S. Chaudhuri, M. Vardi: Optimization of real conjunctive queries. In: [PODS93], pp. 59–70.

[CJMV95] P. Constantopoulos, M. Jarke, J. Mylopoulos, Y. Vassiliou: The software information base: A server for reuse. VLDB Journal, **4**(1), pp. 1–43, 1995.

[CKL*97] L. Colby, A. Kawaguchi, D. Lieuwen, I.S. Mumick, K. Ross: Supporting multiple view maintenance policies. In: [SIGM97], pp. 405–416.

[CKPS95] S. Chaudhuri, S. Krishnamurthy, S. Potamianos, K. Shim: Optimizing queries with materialized views. In: Proceedings of the 11th International Conference on Data Engineering (ICDE), Taipei, pp. 190–200. IEEE Computer Society, March 1995.

[CoHS91] C. Collet, M.N. Huhns, W. Shen: Resource integration using a large knowledge base in Carnot. IEEE Computer, **24**(12), pp. 55–62, 1991.

[CoNS99] S. Cohen, W. Nutt, A. Serebrenik: Rewriting aggregate queries using views. In: Proceedings of the 18th Symposium on the Principles of Database Systems (PODS 99), Philadelphia, PA. 1999.

[Coll96] G. Colliat: OLAP, relational, and multidimensional database systems. SIGMOD Record, **25**(3), September 1996.

[CRGW96] S.Chawathe, A. Rajaraman, H. Garcia-Molina, J. Widom: Change detection in hierarchically structured information. In: [SIGM96], pp. 493–504.

[Davi73] M. Davis: Hilbert's tenth problem is unsolvable. American Mathematical Monthly, **80**, 1973.

[Day93] R.G. Day: Quality Function Deployment: Linking a Company with Its Customers. ASQC Quality Press, Milwaukee, WI, 1993.

[Daya87] U. Dayal: Of nests and trees: a unified approach to processing queries that contain nested subqueries, aggregates, and quantifiers. In: [VLDB87], pp. 197–208.

[Dean97] E.B. Dean: Quality functional deployment from the perspective of competitive advantage. http://mijuno.larc.nasa.gov/dfc/qfd.html, 1997.

[DeDM96] G. De Giacomo, F. Donini, F. Massaçci: Exptime tableaux for ALC. In: Proceedings of the International Workshop on Description Logics, Cambridge, MA. AAAI Press/ MIT Press, 1996.

[Deje95] E.X. Dejesus: Dimensions of data. BYTE, April 1995.

[DeLe95] G. De Giacomo, M. Lenzerini: What's in an aggregate: Foundations for description logics with tuples and sets. In: Proceedings of the 14th International Joint Conference on Artificial Intelligence (IJCAI), Montreal, Canada, pp. 801–807. Morgan Kaufmann, 1995.

[DeLe96] G. De Giacomo, M. Lenzerini: TBox and ABox reasoning in expressive description logics. In: Proceedings of the 5th International Conference on the Principles of Knowledge Representation and Reasoning (KR), Cambridge, MA, pp. 316–327. Morgan Kaufmann, 1996.

[Dema97] M. Demarest. The politics of data warehousing. Available at http://www.hevanet.com/demarest/marc/dwpol.html.

[DeMc92] W.H. Delone, E.R. McLean: Information systems success: The quest for the dependent variable. Information Systems Research, **3**(1), 1992.

[DeNa96] G. De Giacomo, P. Naggar: Conceptual data model with structured objects for statistical databases. In: Proceedings of 8th International Conference on Statistical Database Management Systems (SSDBM), Stockholm, Sweden, pp. 168–175. IEEE Computer Society Press, 1996.

[DEXA98] Proceedings of the 9th International Workshop on Database and Expert Systems Applications (DEXA). IEEE Computer Society, Vienna, Austria, August 1998.

[DiWu91] D.M. Dilts, W. Wu: Using knowledge-based technology to integrate CIM databases. IEEE Transactions on Knowledge and Data Engineering, 3(2), pp. 237–245, 1991.

[DJLS96] S. Dar, H.V. Jagadish, A.Y. Levy, D. Srivastava: Answering queries with aggregation using views. In: [VLDB96], pp. 318–329.

[DWQ99] DWQ Consortium: Homepage of the ESPRIT project, Foundations of Data Warehouse Quality (DWQ), available at http://www.dbnet.ece.ntua.gr/~dwq/, 1999.

[Dyre96] C. Dyreson: Information retrieval from an incomplete data cube. In: [VLDB96], pp. 532–543.

[EDBT96] Proceedings of the 5th International Conference on Extending Database Technology (EDBT), Avignon, France, LNCS 1057. Springer, 1996.

[EDBT98] Proceedings of the 6th International Conference on Extending Database Technology (EDBT), Valencia, Spain, LNCS 1377. Springer, 1998.

[EDD98] Home Page of EDD Data Cleanser tool, available at http://hood.npsa.com/edd, 1998.

[Engl99] L.P. English: Improving Data Warehouse and Business Information Quality. Wiley, 1999.

[ETI98] Home Page of ETI (Evolutionary Technologies International), available at http://www.evtech.com, 1998.

[Fink82] S. Finkelstein: Common expressions analysis in database applications. In: Proceedings of the ACM SIGMOD International Conference on Management of Data, Orlando, FL, pp. 235–245. ACM Press, June 1982.

[Fink95] R. Finkelstein: Multidimensional databases: Where relational fears to tread. Database Programming and Design, April 1995.

[Fink96] R. Finkelstein: Understanding the Need for Online Analytical Servers. White Paper from Arbor Software, Inc., 1996.

[FrSa98] E. Franconi, U. Sattler: Reasoning with multidimensional aggregation: a preliminary report. DWQ Technical Report, 1998.

[FrSa99] E. Franconi, U. Sattler: A data warehouse conceptual data model for multidimensional aggregation. In: Proceedings of the International Workshop Design and Management of Data Warehouses (DMDW '99), Heidelberg, Germany. 1999.

[GaLY98] H. Garcia-Molina., W. Labio, J. Yang: Expiring data in a warehouse. In: Proceedings of the 24th International Conference on Very Large Data Bases (VLDB), New York, pp. 500–511. Morgan Kaufmann, 1998.

[Garv88] D.A. Garvin: Managing Quality: The Strategic and Competitive Edge. N.Y. Free Press, 1988.

[GaSC95] M. Garcia-Solaco, F. Saltor, M. Castellanos: A semantic-discriminated approach to integration in federated databases. In: Proceedings of the 3rd International Conference on Cooperative Information Systems (CoopIS), Vienna, Austria, pp. 19–31. 1995.

[GaSC95a] M. Garcia-Solaco, F. Saltor, M. Castellanos: A structure based schema integration methodology. In: Proceedings of the 11th International Conference on Data Engineering (ICDE), Taipei, Taiwan, pp. 505–512. IEEE Computer Society, 1995.

[GBBI96] A. Goñi, J. Bermúdez, J. M. Blanco, A. Illarramendi: Using reasoning of description logics for query processing in multidatabase systems. In: Working notes of the ECAI-96 Workshop on Knowledge Representation Meets Databases (KRDB), Budapest, Hungary. 96.

[GBLP96] J. Gray, A. Bosworth, A. Layman, H. Pirahesh: Data cube: A relational aggregation operator generalizing group-by, cross-tabs, and sub-totals. In: Proceedings of the 12th International Conference on Data Engineering (ICDE), New Orleans, pp. 152–159. IEEE Computer Society, February 1996. Also Microsoft Technical Report, MSR-TR-95-22, available at http://www.research.microsoft. com/ ~gray.

[GBLP97] J. Gray, A. Bosworth, A. Layman, H. Pirahesh: Data cube: A relational operator generalizing group-by, cross-tabs, and sub-totals. Data Mining and Knowledge Discovery **1**(1), pp. 29–53, 1997.

[GeJJ97] M. Gebhardt, M Jarke, S. Jacobs: A toolkit for negotiation support on multi-dimensional data. In: [SIGM97], pp. 348–356.

[GFSS99] H. Galhardas, D. Florescu, D. Shasha and E. Simon: An Extensible Framework for Data Cleaning,Technical Report 3742, INRIA 1999.

[GFSS00] H. Galhardas, D. Florescu, D. Shasha and E. Simon. Ajax: An Extensible Data Cleaning Tool. In: Proceedings of the ACM SIGMOD International Conference on the Management of Data, pp. 590, Dallas, Texas.ACM Press, 2000.

[GHRU97] H. Gupta, V. Harinarayan, A. Rajaraman, J. Ullman: Index selection for OLAP. In: Proceedings of the 13th International Conference on Data Engineering (ICDE), Birmingham, UK, pp. 208–219. IEEE Computer Society, April 1997.

[GoLa01] J. Goldstein, P.-A. Larson: optimizing queries using materialized views – a practical, scalable solution. In: Proceedings of the ACM-SIGMOD International Conference on Management of Data, Santa Barbara, CA, pp. 331–342.ACM Press 2001.

[GoLN92] W. Gotthard, P.C. Lockemann, A. Neufeld: System-guided view integration for object-oriented databases. IEEE Transactions on Knowledge and Data Engineering, **4**(1), pp. 1–22, 1992.

[GPC*92] J. Geller, Y. Perl, P. Cannata, A.P. Sheth, E. Neuhold: A case study of structural integration. In: Proceedings of the 1st International Conference on Information and Knowledge Management (CIKM), Baltimore, MD, pp. 102–111. 1992.

[GPNS92] J. Geller, Y. Perl, E. Neuhold, A. Sheth: Structural schema integration with full and partial correspondence using the dual model. Information Systems, **17**(6), pp. 443–464, 1992.

[Gree97] L. Greenfield: Data Warehousing Information Center. http://pwp. starnetinc.com/larryg/index.html, 1997.

[GrLi95] T.Griffin, L. Libkin: Incremental maintenance of views with duplicates. In: [SIGM95], pp. 328–339.

[GuHQ95] A. Gupta, V. Harinarayan, D. Quass: Aggregate query processing in data warehouses. In: [VLDB95], pp. 358–369.

[GuJM96] A. Gupta, H.V. Jagadish, I.S. Mumick: Data integration using self-maintainable views. In: [EDBT96], pp. 140–144.

[GuMS93] A. Gupta, I.S. Mumick, V.S. Subrahmanian: Maintaining views incrementally. In: Proceedings of ACM SIGMOD International Conference on Management of Data, Washington DC, pp. 157–166. ACM Press, 1993.

[GuMu95] A. Gupta, I.S. Mumick: Maintenance of materialized views: Problems, techniques, and applications. IEEE Data Engineering Bulletin, Special Issue on Materialized Views and Data Warehousing, **18**(2), pp. 3–19, 1995.

[Gupt97] H. Gupta: Selection of views to materialize in a data warehouse. In: [ICDT97], pp. 98–112.

[GyLa97] M. Gyssens, L.V.S. Lakshmanan: A foundation for multi-dimensional databases. In: [VLDB97], pp. 106–115.

[HaCl88] J.R. Hauser, D. Clausing: The house of quality. The Harvard Business Review, May–June, No. 3, 1988.

[Hans87] E. Hanson: A performance analysis of view materialization strategies. In: Proceedings of the ACM SIGMOD International Conference on Management of Data, San Francisco, CA, pp. 440–453. ACM Press, 1987.

[HaRU96] V. Harinarayan, A. Rajaraman, J. Ullman: Implementing data cubes efficiently. In: [SIGM96], pp. 205–227.

[HBG*97] J. Hammer, M. Breuning, H. Garcia-Molina, S. Nestorov, V. Vassalos, R. Yerneni: Template-based wrappers in the TSIMMIS system. In: [SIGM97], pp. 532–535.

[HeSt95] M.A. Hernandez, S.J. Stolfo: The merge/purge problem for large databases. In: [SIGM95], pp. 127–138.

[HeSt98] M.A. Hernandez, S.J. Stolfo: Real-world data is dirty: Data cleansing and the merge/purge problem. Journal of Data Mining and Knowledge Discovery, **2**(1), pp. 9–37, 1998.

[HGW*95] J. Hammer, H. Garcia-Molina, J. Widom, W. Labio, Y. Zhuge: The Stanford data warehousing project. IEEE Data Engineering Bulletin, Special Issue on Materialized Views and Data Warehousing, **18**(2), pp. 41–48, 1995.

[HJK*93] M.N. Huhns, N. Jacobs, T. Ksiezyk, W. Shen, M.P. Singh, P.E. Cannata: Integrating enterprise information models in Carnot. In: Proceedings of the International Conference on Cooperative Information Systems (CoopIS), Rotterdam, The Netherlands, pp. 32–42. IEEE Computer Society, 1993.

[HKRW90] Y.U. Huh, F.R. Keller, T.C. Redman, A.R. Watkins: Data quality. Information and Software Technology, **32**(8), 1990.

[HMM*78] D. Halloran, S.J. Manchester, J.W. Moriarty, R.E. Riley, J.R. Rohrman, T.B. Skramstad: Systems development quality control. MIS Quarterly, **2**(4), pp. 1–14, 1978.

[HNSS95] P.J. Haas, J.F. Naughton, S. Seshadri, A.N. Swami: Sampling-based estimation of the number of distinct values of an attribute. In: [VLDB95], pp. 311–322.

[Horr99] I. Horrocks: FaCT and iFaCT. In: Proceedings of the International Workshop on Description Logics (DL'99), CEUR-WS 22, Linköping, Sweden, pp. 133–135. 1999.

[Hull86] R. Hull: Relative information capacity of simple relational database schema. SIAM Journal of Computing, **15**(3), pp. 856–886, 1986.

[Hull97] R. Hull: Managing semantic heterogeneity in databases: A theoretical perspective. In: Proceedings of the 16th ACM SIGACT SIGMOD SIGART Symposium on Principles of Database Systems (PODS), Tucson, AZ, pp. 51–61. ACM Press, 1997.

[Hurw97] M. Hurwicz: Take your data to the cleaners. BYTE, January 1997.

[Huyn96] N. Huyn: Efficient view self-maintenance. Full version available at http://www-db.stanford-edu/warehousing/html/publications.html, 1996.

[Huyn97] N. Huyn: Multiple-view self-maintenance in data warehousing environments. In: [VLDB97], pp. 26–35.

[HuZh96] R. Hull, G. Zhou: A framework for supporting data integration using the materialized and virtual approaches. In: [SIGM96], pp. 481–492.

[HuZh96] R. Hull, G. Zhou: Towards the study of performance trade-off between materialized and virtual integrated views. In: [MuGu96], pp. 91–102.

[HyRo96] L. Hyatt, L. Rosenberg: A software quality model and metrics for identifying project risks and assessing software quality. In: Proceedings of the 8th Annual Software Technology Conference, Utah. 1996.

[ICDT97] Proceedings of the International Conference on Database Theory (ICDT), Delphi, Greece, LNCS 1186. Springer, 1997.

[IEEE97] IEEE: Information Technology. http://standards.ieee.org/catalog/ it. html, 1997.

[IiKo87] J. Iivari, E. Koskela: The PIOCO model for information systems design. MIS Quarterly, **11**(3), pp. 401–419, 1987.

[Info97] Informix Inc.: The INFORMIX-MetaCube product suite. http://www. informix.com/informix/products/new_plo/metabro/metabro2.htm, 1997.

[Inmo96] W. H. Inmon: Building the Data Warehouse, 2nd edition. Wiley, 1996.

[Inmo97] B. Inmon. The Data Warehouse Budget. DM Review, Jan. 1997. http://www.dmreview.com/master.cfm?NavID=55&EdID=1315.

[ISIA97] C. Imhoff, R. Sousa, W.H. Inmon, C. Amhoff: Corporate Information Factory. Wiley, 1997.

[ISO92] ISO: ISO 9000 International Standards for Quality Management, Geneva: International Organization for Standardization, 1992.

[ISO97] ISO: International Organization for Standardization, http://www.iso. Ch, 1997.

[Jahn96] M. Jahnke: Technical report. Sabre Decision Technologies, 4255 Amon Carter Blvd., MD4364, Fort Worth, TX 76155, USA.

[Jans88] M. Janson: Data quality: The Achilles heel of end-user computing. Omega: International Journal of Management Science, **16**(5), 1988.

[JaPo92] M. Jarke, K. Pohl: Information systems quality and quality information systems. In: Proceedings of the IFIP Working Conference on the Impact of Computer Supported Technologies on Information Systems Development, Minneapolis, MI, pp. 345–375. North-Holland, 1992.

[JaRo88] M. Jarke, T. Rose: Managing knowledge about information system evolution. In: Proceedings of the ACM SIGMOD International Conference of the Management of Data, Chicago, IL, pp. 303–311. ACM Press1988.

[JaVa97] M. Jarke, M. Vassiliou: Foundations of data warehouse quality: An overview of the DWQ project. In: Proceedings of the 2nd International Conference on Information Quality, Cambridge, MA, pp. 299–313. 1997.

[JeQJ98] M.A. Jeusfeld, C. Quix, M. Jarke: Design and analysis of quality information for data warehouses. In: Proceedings of the 17th International Conference on Conceptual Modeling (ER), Singapore, LNCS 1507, pp. 349–362. Springer, 1998.

[Jeus92] M.A. Jeusfeld: Update Control in Deductive Object Bases. PhD thesis, University of Passau (in German), 1992.

[JGJ*95] M. Jarke, R. Gallersdörfer, M. Jeusfeld, M. Staudt, S. Eherer: ConceptBase – a deductive object base for meta data management. Journal of Intelligent Information Systems, **4**(2), pp. 167–192, 1995.

[JJQV99] M. Jarke, M.A. Jeusfeld, C. Quix, P. Vassiliadis: Architecture and quality in data warehouses: An extended repository approach. Information Systems, **24**(3), pp. 229–253, 1999 (Special Issue on Best Papers from CAiSE 98, Pisa, Italy).

[JQB*99] M. Jarke, C. Quix, G. Blees, D. Lehmann, G. Michalk, S. Stierl: Improving OLTP data quality using data warehouse mechanisms. In: Proceedings of the ACM-SIGMOD Conference on Management of Data, Philadelphia, PA.ACM Press 1999.

[JQC*00] M. Jarke, C. Quix, D. Calvanese, M. Lenzerini, M. Franconi, P. Ligoudistianos, P. Vassiliadis, Y. Vassiliou: Concept-based design of data warehouses–the DWQ demonstrators (Demonstration). In: W. Chen, J. Naughton, P.A. Bernstein (eds.) Proceedings of the ACM SIGMOD International Conference on Management of Data, Dallas, TX, USA, 14–19 May, p. 591. 2000.

[Joha94] P. Johanneson: Schema standardization as an aid in view integration. Information Systems, **19**(3), pp. 275–290, 1994.

[JoKl83] D.S. Johnson, A. Klug: Optimizing conjunctive queries that contain untyped variables. SIAM Journal on Computing, **12**(4), pp. 616–640, 1983.

[Jürg02] M. Jürgens: Index Structures for Data Warehouses. LNCS 1859, Springer, 2002.

[Jura88] J.M. Juran (ed): Quality Control Handbook. McGraw-Hill, 1988.

[Jura92] J.M. Juran: Juran on Quality Design: The New Steps for Planning Quality into Goods and Services. Free Press, 1992.

[Kack86] R. Kackar: Taguchi's quality philosophy: Analysis and commentary. Quality Progress, December 1986.

[KaWa96] G. Kamp, H. Wache: CTL – a description logic with expressive concrete domains. Technical report LKI-M-96/01, Labor für Künstliche Intelligenz, Universität Hamburg, Germany, 1996.

[Kena95] Kenan Technologies: An introduction to multidimensional database technology. http://www.kenan.com/acumate/mddb_toc.htm, 1995.

[KiEp83] W. King, B.J. Epstein: Assessing information system value: An experiment study. Decision Sciences, **14**(1), 1983.

[Kimb96] R. Kimball: The Data Warehouse Toolkit. Wiley, 1996.

[KiRS97] J.-U. Kietz, U. Reimer, M. Staudt: Mining insurance data at Swiss Life. In: [VLDB97], pp. 562–566.

[KLM*97] A. Kawaguchi, D. Lieuwen, I.S. Mumick, D. Quass, K. Ross: Concurrency control theory for deferred materialized views. In: [ICDT97], pp. 306–320.

[KLSS95] T. Kirk, A.Y. Levy, Y. Sagiv, D. Srivastava: The Information Manifold. In: Proceedings of the AAAI 1995 Spring Symposium on Information Gathering from Heterogeneous, Distributed Environments, pp. 85–91. 1995.

[KoLo98] A. Koschel, P. Lockemann: Distributed events in active database systems – Letting the genie out of the bottle. Data and Knowledge Engineering, **25**(1-2), pp. 11–28, 1998.

[KoRo98] Y. Kotidis, N. Roussopoulos: An alternative storage organization for ROLAP aggregate views based on cubetrees. In: Proceedings of the 1998 ACM SIGMOD International Conference on Management of Data, Seattle, WA, pp. 249–258. ACM Press, 1998.

[Krie79] C. Kriebel: Evaluating the quality of information system. In: N. Szysperski, E. Grochla (eds): Design and Implementation of Computer Based Information Systems. Sijthoff and Noordhoff, Germantown, 1979.

[KrLK91] R. Krishnamurthy, W. Litwin, W. Kent: Language features for interoperability of databases with schematic discrepancies. In: Proceedings of the ACM SIGMOD International Conference of the Management of Data, Denver, CO, pp. 40–49. ACM Press, 1991.

[KRRT98] R. Kimball, L. Reeves, M. Ross, W. Thornthwaite: The Data Warehouse Lifecycle Toolkit: Export Methods for Designing, Developing, and Deploying Data Warehouses. Wiley, 1998.

[LaGa96] W. Labio, H. Garcia-Molina: Efficient snapshot differential algorithms for data warehousing. In: [VLDB96], pp. 63–74.

[Laud86] K.C. Laudon: Data quality and due process in large interorganizational record systems. Communications of the ACM, **29**(1), pp. 4–11, 1986.

[LeAW98] W. Lehner, J. Albrecht, H. Wedekind: Normal forms for multidimensional databases. In: [SSDB98], pp. 63–72.

[LeGu90] D. Lenat, R. V. Guha: Building large knowledge-based systems: Representation and inference in the Cyc Project. Addison Wesley, Reading, MA, 1990.

[Lehn98] W. Lehner: Modeling large scale OLAP scenarios. In: [EDBT98], pp. 153–167.

[LeMu96] A.Y. Levy, I.S. Mumick: Reasoning with aggregation constraints. In: [EDBT96], pp. 514–534.

[LeRO96] A.Y. Levy, A. Rajaraman, J.J. Ordille: Query answering algorithms for information agents. In: Proceedings of the 13th National Conference on Artificial Intelligence (AAAI), Portland, OR, pp. 40–47. AAAI Press/MIT Press, 1996.

[LeSa95] A.Y. Levy, Y. Sagiv: Semantic query optimization in datalog programs. In: [PODS95], pp. 163–173.

[LeSK95] A.Y. Levy, D. Srivastava, T. Kirk: Data model and query evaluation in global information systems. Journal of Intelligent Information Systems, 5(2), pp. 121–143, 1995.

[LiWa96] C. Li, X. Sean Wang: A data model for supporting online analytical processing. In: Proceedings of the 5th International Conference on Information and Knowledge Management (CIKM), Rockville, MD, pp. 81–88. ACM Press, 1996.

[LMSS93] A.Y. Levy, I.S. Mumick, Y. Sagiv, O. Shmueli: Equivalence, query-reachability, and satisfiability in datalog extensions. In: [PODS93], pp. 109–122.

[LMSS95] A.Y. Levy, A.O. Mendelzon, Y. Sagiv, D. Srivastava: Answering queries using views. In: [PODS95], pp. 95–104.

[MBJK90] J. Mylopoulos, A. Borgida, M. Jarke, M. Koubarakis: Telos: A language for representing knowledge about information systems. ACM Transactions on Information Systems, 8(4), pp. 325–362, 1990.

[McRW78] J.A. McCall, P.K. Richards, G.F. Walters: Factors in software quality. Technical report, Rome Air Development Center, 1978.

[MeCo96] Metadata Coalition: Proposal for version 1.0 metadata interchange specification. http://www.metadata.org/standards/toc.html, July 1996.

[MeDC99] MetaData Coalition. Open Information Model, version 1.0 (1999). Available http://at www.MDCinfo.com.

[MiAk94] S. Mizuno, Y. Akao (ed): QFD: The Customer-Driven Approach to Quality Planning and Development. Asian Productivity Organization, Tokyo, Japan, available from Quality Resources, One Water Street, White Plains, NY, 1994.

[Micr96] Microsoft: Microsoft SQL Server and data warehousing. Presentation, available at http://www.microsoft.com:80/sql/datawarehouse/ dw_pp/ index.htm, 1996.

[MiPE94] Microsoft Professional Editions: ODBC 2.0 Programmer's Reference and SDK Guide. Microsoft Press, Redmond, WA, 1994.

[MiYR94] R.J. Miller, Y.E. Yoannidis, R. Ramakrishnan: Schema equivalence in heterogeneous systems: Bridging theory and practice. Information Systems, 19(1), pp. 3–31, 1994.

[MoEl96] A.E. Monge, C.P. Elkan: The field matching problem: Algorithms and applications. In: Proceedings of Knowledge Discovery and Data Mining Conference (KDD), Portland, OR, pp. 267–270. AAAI Press, 1996.

[More82] R.C. Morey: Estimating and improving the quality of information in a MIS. Communications of the ACM, 25(5), pp. 337–342, 1982.

[MStr95] MicroStrategy, Inc.: Relational OLAP: An enterprise-wide data delivery architecture. White Paper, available at http://www.strategy.com/wp_a_i1.htm, 1995.

[MStr97] MicroStrategy, Inc.: MicroStrategy's 4.0 Product Line. http://www.strategy.com/launch/4_0_arc1.htm, 1997.

[Muck96] H. Mucksch: Characteristics, components and organizational structures of data warehouse (in German). In: H. Mucksch, W. Behme (eds): The Data Warehouse Concept. Gabler Verlag, 1996.

[MuGu96] I.S. Mumick, A. Gupta (eds.): Materialized views: techniques and applications (VIEWS 1996). In: Proceedings International Workshop held in cooperation with ACM SIGMOD, Montreal, Canada. 1996.

[MuQM97] I.S. Mumick, D. Quass, B.S. Mumick: Maintenance of data cubes and summary tables in a warehouse. In: [SIGM96], pp. 100–111.

[MuSh95] I.S. Mumick, O. Shmueli: How expressive is stratified aggregation. Annals of Mathematics and Artificial Intelligence, **15**(3–4), 1995.

[MyCN92] J. Mylopoulos, L. Chung, B. Nixon: Representing and using nonfunctional requirements: a process-centered approach. IEEE Transactions on Software Engineering, **18**(6), pp. 483–497, 1992.

[NiJa98] H.W. Nissen, M. Jarke: Requirements engineering repositories: Formal support for informal teamwork methods. In: M. Broy (ed.): Requirements Targeting Software and Systems, LNCS 1526. Springer, 1998 (extended version in Information Systems, Special Issue on Meta Modeling and Method Engineering, **24**(2), pp. 131–158, 1999).

[NiJa99] M. Nicola, M. Jarke: Increasing the expressiveness of analytical performance models for replicated databases. In: Proceedings of the International Conference on Database Theory (ICDT), Jerusalem, Israel, LNCS 1540, pp. 131–149. Springer, 1999.

[NiJa00] M. Nicola, M. Jarke: Performance modeling of distributed and replicated databases. Research survey, IEEE Transactions on Knowledge and Data Engineering **12**(4), pp. 645–672, 2000.

[NuSS98] W. Nutt, Y. Sagiv, S. Shurin: Deciding equivalences among aggregate queries. In: [PODS98], pp. 214–223.

[OiBa92] M. Oivo, V. Basili: Representing software engineering models: the TAME goal-oriented approach. IEEE Transactions on Software Engineering, **18**(10), pp. 886–898, 1992.

[OlCo95] Olap Council: The OLAP glossary. http://www.olapcouncil.org/ research/glossaryly.htm, January 1995.

[OlRo90] F. Olken, D. Rotem: Random sampling from database files: a survey. In: Proceedings of the 5th International Conference on Statistical and Scientific Database Management (SSDBM), Charlotte, NC, pp. 92-111. 1990.

[ONGr95] P. O'Neil, G. Graefe: Multi-table joins through bitmapped join indexes. SIGMOD Record, 24(3), pp. 8–11, 1995.

[ONQu97] P. O'Neil, D. Quass: Improved query performance with variant indexes. In: [SIGM97], pp. 38–49.

[OrCo96] Oracle Corporation: Oracle server application developer's guide, 1996.

[OrCo96a] Oracle Corporation: Oracle7 server utilities user's guide, SQL* Loader Concepts, available at http://marshal.uwyo.edu/Oracle/ DOC/ server/doc/SUT73/ch3.htm#toc058, 1996.

[OrCo97] Oracle Corporation: Oracle Express Server. http://www.oracle.com/ products/olap/ html/oes.html, 1997.

[Orli97] R.J. Orli: Data extraction, transformation, and migration tools. Kismet Analytic Corp, http://www.kismeta.com/ex2.htm, 1997.

[Orr98] K. Orr. Data quality and systems theory. Communications of the ACM, **41**(2), pp. 66–71, 1998.

[OuNa94] A.M. Ouksel, C. F. Naiman: Coordinating context building in heterogeneous information systems. Journal of Intelligent Information Systems, **3**(2), pp. 151–183, 1994.

[OzOM85] G. Özsoyoglu, Z.M. Özsoyoglu, F. Mata: A language and a physical organization technique for summary tables. In: Proceedings of the ACM SIGMOD International Conference on Management of Data, Austin, TX, pp. 3–16. ACM Press, 1985.

[OzOM87] G. Özsoyoglu, Z.M. Özsoyoglu, V. Matos: Extending relational algebra and relational calculus with set-valued attributes and aggregation functions. ACM Transactions on Database Systems, **12**(4), pp. 566–592, 1987.

[OzVa99] M. T. Ozsu, P. Valduriez. Principles of Distributed Database Systems. Prentice Hall, 1999.

[PaRe90] R.W. Pautke, T.C. Redman: Techniques to control and improve quality of data large databases. In: Proceedings of Statistics Canada Symposium. 1990.

[PeCr00] N. Pendse, R. Creeth: The OLAP report. Business Intelligence Inc., London 2000.

[Pelt91] C. Peltason: The BACK system – an overview. SIGART Bulletin, 2(3), pp. 114–119, 1991.

[PMB*91] P.F. Patel-Schneider, D.L. McGuiness, R.J. Brachman, L. Alperin Resnick, A. Borgida: The CLASSIC knowledge representation system: Guiding principles and implementation rational. SIGART Bulletin, 2(3), pp. 108–113, 1991.

[PODS93] Proceedings of the 12th ACM SIGACT SIGMOD SIGART Symposium on Principles of Database Systems (PODS), Washington DC. ACM Press, May 1993.

[PODS95] Proceedings of the 14th ACM SIGACT SIGMOD SIGART Symposium on Principles of Database Systems (PODS), San Jose, CA. ACM Press, 1995.

[PODS98] Proceedings of 17th ACM SIGACT SIGMOD SIGART Symposium on Principles of Database Systems (PODS), Seattle, WA. ACM Press, 1998.

[PuIn98] Home Page of Carleton's Pure Integrator tool, available at http://www.apertus.com/products/Integrate/index.htm, 1998.

[QGMW96] D.Quass, A. Gupta, I.S. Mumick, J. Widom: Making views self-maintainable for data warehousing. In: Proceedings of the Conference on Parallel and Distributed Information Systems, Miami, FL, pp. 158–169. 1996.

[Qian96] X. Qian: Correct schema transformations. In [EDBT98], pp. 114–128.

[QiWi91] X. Qian, G. Wiederhold: Incremental recomputation of active relational expressions. IEEE Transactions on Knowledge and Data Engineering, 3(3), pp. 337–341, September 1991.

[QuWi97] D. Quass, J. Widom: On-line warehouse view maintenance. In: [SIGM97], pp. 393–404.

[Rade95] N. Raden: Data, data everywhere. Information Week, October, 1995.

[Rade96] N. Raden: Technology tutorial – Modeling a data warehouse. Information Week, January 29, 1996.

[RaHa00] E. Rahm, H. Do. Data Cleaning: Problems and Current Approaches. Bulletin of the Technical Committee on Data Engineering, **23**(4), pp. 3–13, December 2000.

[RaHe00] V. Raman, J. Hellerstein. Potters Wheel: An Interactive Framework for Data Cleaning and Transformation. Technical Report University of California at Berkeley, Computer Science Division, 2000. http://www.cs.berkeley.edu/~rshankar/papers/pwheel.pdf.

[RaRi91] M. Rafanelli, F.L. Ricci: A functional model for macro-databases. SIGMOD Record, **20**(1), March 1991.

[RBSI96] Red Brick Systems, Inc.: Data warehouse applications, 1996.

[RBSI97] Red Brick Systems, Inc.: Red Brick Warehouse 5.0. http://www.redbrick.com/rbs-g/html/whouse50.html, 1997.

[Redm92] T.C. Redman: Data Quality: Management and Technology, N.Y. Bantam Books, 1992.

[RoKR97] N. Roussopoulos, Y. Kotidis, M. Roussopoulos: Cubetree: Organization of and bulk updates on the data cube. In: [SIGM97], pp. 89–99.

[RoSS96] K.A. Ross, D. Srivastava, S. Sudarshan: Materialized view maintenance and integrity constraints: Trading space for time. In: [SIGM96], pp. 447–458.

[RPRG94] M.P. Reddy, B.E. Prasad, P.G. Reddy, A. Gupta: A methodology for integration of heterogeneous databases. IEEE Transactions on Knowledge and Data Engineering, **6**(6), pp. 920–933, 1994.

[Rudl96] D. Rudloff: Terminological reasoning and conceptual modeling for data warehouse. In: Working Notes of the ECAI-96 Workshop on Knowledge Representation Meets Databases (KRDB), 1996.

[Rund99] E. Rundensteiner (ed.). Special Issue on Data Transformations. Bulletin of the IEEE Technical Committee on Data Engineering, **22**(1), March 1999.

[Sach97] S. Sachdeva: Metadata for data warehouse. White Paper, Sybase, Inc. http://www.sybase.com/services/dwpractice/meta.html, 1997.

[Sahi96] K. Sahin: Multidimensional database technology and data warehousing, http://www.kenan.com/acumate/byln_mdw.htm, 1996.

[SaSa92] Y. Sagiv, Y. Saraiya: Minimizing restricted-fanout queries. Discrete Applied Mathematics, 40, pp. 245–264, 1992.

[SaYa80] Y. Sagiv, M. Yannakakis: Equivalence among relational expressions with the union and difference operators, Journal of the ACM, **27**(4), pp. 633–655, 1980.

[ScBJ00] E. Schäfer, J.-D. Becker, M. Jarke: DB-Prism – integrated data warehouses and knowledge networks for bank controlling. Industrial paper. In: Proceedings of the 26th International Conference on Very Large Data Bases (VLDB 2000), Cairo, Egypt, pp. 715–718. Morgan Kaufmann 2000.

[Schn90] N. Schneidewind: Standard for a software quality metrics methodology. Software Engineering Standards Subcommittee of the IEEE, 1990.

[SCJL98] J. Smith, V. Castelli, A. Jhingran, C. Li: Dynamic assembly of views in data cubes. In: [PODS98], pp. 274–283.

[SDJL96] D. Srivastava, S. Dar, H.V. Jagadish, A.Y. Levy: Answering queries with aggregation using views. In: [VLDB96], pp. 318–329.

[SDNR96] A. Shukla, P.M. Deshpande, J.F. Naughton, K. Ramasamy: Storage estimation for multidimensional aggregates in the presence of hierarchies. In: [VLDB96], pp. 522–531.

[Sell88] T. Sellis: Intelligent caching and indexing techniques for relational database systems. Information Systems, **13**(2), pp. 175–188, 1988.

[SePa89] A. Segev, J. Park: Maintaining materialized views in distributed databases. In: Proceedings of the 5th International Conference on Data Engineering (ICDE), Los Angeles, CA, pp. 262–270. IEEE Computer Society, 1989.

[Shas98] Dennis Shasha: Personal communication, 1998.

[ShGN93] A.P. Sheth, S.K. Gala, S.B. Navathe: On automatic reasoning for schema integration. Journal of Intelligent and Cooperative Information Systems, **2**(1), pp. 23–50, 1993.

[Shos97] A. Shoshani: OLAP and statistical databases: Similarities and differences. In: Tutorials of the 16th ACM SIGACT SIGMOD SIGART Symposium on Principles of Database Systems (PODS), 1997.

[ShTy98] C. Shilakes, J. Tylman. Enterprise Information Portals. Enterprise Software Team. November 1998. Available at http://www.sagemaker.com/company/downloads/eip/indepth.pdf.

[SIGM95] Proceedings of ACM SIGMOD International Conference on Management of Data, San Jose, CA. ACM Press, 1995.

[SIGM96] Proceedings of ACM SIGMOD International Conference on Management of Data, Montreal, Canada. ACM Press, 1996.

[SIGM97] Proceedings of ACM SIGMOD International Conference on Management of Data. Tucson, AZ. ACM Press, 1997.

[SiKo95] E. Simon, A. Kotz-Dittrich: Promises and realities of active database systems. In: [VLDB95], pp. 642–653.

[Simo98] A.R. Simon: 90 Days to the Data Mart. Wiley, 1998.

[SJGP90] M. Stonebraker, A. Jhingran, J. Goh, S. Potamianos: On rules, procedures, caching and views in database systems. In: Proceedings of the ACM SIGMOD International Conference on Management of Data, Atlantic City, NJ, pp. 245–254. ACM Press, May 1990.

[SoAG96] Software AG: SourcePoint White Paper. Software AG, Uhlandstr. 12, D-64297 Darmstadt, Germany, 1996.

[SpPD92] S. Spaccapietra, C. Parent, Y. Dupont: Model independent assertions for integration of heterogeneous schemas. VLDB Journal, **1**(1), pp. 81–126, 1992.

[SSDB98] Proceedings of 10th International Conference on Scientific and Statistical Database Management (SSDBM), Capri, Italy. IEEE Computer Society, July 1998.

[STGI96] Stanford Technology Group, Inc.: Designing the data warehouse on relational databases. http://www.informix.com/informix/corpinfo/zines/whitpprs/stg/metacube.htm, 1996.

[StJa96] M. Staudt, M. Jarke: Incremental maintenance of externally materialized views. In: [VLDB96], pp. 75–86.

[StJa00] M. Staudt, M. Jarke: View management support in advanced knowledge base servers. Journal of Intelligent Information Systems **15**(3), pp. 253–285, 2000.

[StKR98] M. Staudt, J.-U. Kietz, U. Reimer: A data mining support environment and its application to insurance data. In: Proceedings of the 4th International Conference on Knowledge Discovery and Data Mining (KDD), New York, pp. 105–111. AAAI Press, August 1998.

[StLW94] D.M. Strong, Y.W. Lee, R.Y. Wang: Beyond accuracy: How organizations are redefining data quality. Technical report, no. TDQM-94-07, Total Data Quality Management Research Program, MIT Sloan School of Management, Cambridge, MA, 1994.

[Svan84] M.I. Svanks: Integrity analysis: Methods for automating data quality assurance. EDP Auditors Foundation, **30**(10), 1984.

[Syba97] Sybase, Inc.: Data warehouse Solutions. http://www.sybase.com/products/dataware/, 1997.

[TAB*97] A. Tomasic, R. Amouroux, P. Bonnet, O. Kapitskaia, H. Naacke, L. Raschid: The distributed information search component (Disco) and the World Wide Web. In: [SIGM97], pp. 546–548.

[ThSe97] D. Theodoratos, T. Sellis: Data warehouse configuration. In: [VLDB97], pp. 126–135.

[ThLS99] D. Theodoratos, S. Ligoudistianos, T. Sellis : Designing the Global Data Warehouse with SPJ queries. In: Proceedings of the 11th International Conference on Advanced Information Systems Engineering (CAiSE'99), LNCS 1626, pp. 180–194. Springer, 1999.

[Thin01] Thinking Networks AG: b2brain – the multi-dimensional information system. http://www.thinking-networks.de/download/ b2brain_e.pdf.

[TrLN99] S. Trisolini, M. Lenzerini, D. Nardi: Data integration and warehousing in Telecom Italia. In: Proceedings of the ACM SIGMOD International. Conference on Management of Data, pp. 538–539.ACM Press 999.

[ToBl88] F. Tompa, J. Blakeley: Maintaining materialized views without accessing base data. Information Systems, **13**(4), pp. 393–406, 1988.

[Ullm97] J.D. Ullman: Information integration using logical views. In: [ICDT97], pp. 19–40.

[VaBQ00] P. Vassiliadis, M. Bouzeghoub, C. Quix. Towards quality-oriented data warehouse usage and evolution. Information Systems, **25**(2), pp. 89–115, 2000.

[VaDM92] R. van der Meyden: The complexity of querying indefinite data about linearly ordered domains. In Proceedings of the 11th ACM SIGACT SIGMOD SIGART Symposium on Principles of Database Systems (PODS), San Diego, CA, pp. 331–345. ACM Press, 1992.

[Vali98] Home Page of Integrity tool available at http://www.vality.com/html/prod-int.html.

[Vass98] P. Vassiliadis: Modeling multidimensional databases, cubes, and cube operations. In: [SSDB98], pp. 53–62.

[Vass00] P. Vassiliadis: Gulliver in the land of data warehousing: practical experiences and observations of a researcher. In: Proceedings of the 2nd International Workshop Design and Maintenance of Data Warehouses (DMDW 00), Stockholm, Sweden, http://CEUR-WS.org/Vol-28/. 2000.

[VeVS99] T. Vetterli, A. Vaduva, M. Staudt: Metadata standards for data warehousing: Open Information Model vs. Common Warehouse Model. SIGMOD Record, 1999.

[ViWi94] V. Vidal, M. Winslett: Preserving update semantics in schema integration. In: Proceedings of the 3rd International Conference on Information and Knowledge Management (CIKM), Gaithersburg, MD, pp. 263–271. ACM Press, 1994.

[VLDB87] Proceedings of the 13th International Conference on Very Large Data Bases (VLDB), Brighton, UK. Morgan Kaufmann, September 1987.

[VLDB95] Proceedings of the 21st International Conference on Very Large Data Bases (VLDB), Zurich, Switzerland. Morgan Kaufmann, August 1995.

[VLDB96] Proceedings of the 22nd International Conference on Very Large Databases (VLDB), Mumbai, India. Morgan Kaufmann, September 1996.

[VLDB97] Proceedings of the 23rd International Conference on Very Large Databases (VLDB), Athens, Greece. Morgan Kaufmann, August 1997.

[VLST01] P. Vassiliadis, S. Ligoudistianos, T. Sellis, D. Theodoratos: Quality-oriented optimization of the design of the global data warehouse. In: Proceedings of the 8th Panhellenic Conference on Informatics, pp. 227–226, Nicosia, Cyprus. 2001.

[VQVJ01] P. Vassiliadis, C. Quix, Y. Vassiliou, M. Jarke: Data warehouse process management. Information Systems **26**(3), 2001, 205–236.

[VVS*01] P.Vassiliadis, Z.Vagena, S.Skiadopoulos, N.Karayannidis, T. Sellis: Arktos–towards the modeling, design, control and execution of ETL processes. Information Systems, **26**(8), pp. 537–561, 2001.

[WaKM93] R.Y. Wang, H.B. Kon, S.E. Madnick: Data quality requirements analysis and modeling. In: Proceedings of the 9th International Conference on Data Engineering (ICDE), Vienna, Austria, pp. 670–677. IEEE Computer Society, 1993.

[WaKo93] R.Y. Wang, H.B. Kon: Towards total data quality management. In: R.Y. Wang, (ed): Information technology action: Trends and perspectives. Prentice Hall, 1993.

[WaMa90] R.Y. Wang, S.E. Madnick: A Polygen model for heterogeneous database systems: The source tagging perspective. Proceedings of the 16th International Conference on Very Large Data Bases (VLDB), Brisbane, Australia, pp. 519–538. Morgan Kaufmann, 1990.

[Wand89] Y. Wand: A proposal for a formal model of objects. In: W. Kim, F. Lochovsky (eds.): Object-Oriented Concepts, Databases, and Applications. ACM Press, 1989.

[Wang98] R.Y. Wang: A product perspective on total data quality management. Communications of the ACM, **41**(2), pp. 58–65, February 1998.

[Wang00] Wang, R. (ed.): Data Quality. Kluwer, 2000.

[WaRG93] R.Y. Wang, M.P. Reddy, A. Gupta: An object-oriented implementation of quality data products. In: Proceedings of the 3rd Annual Workshop on Information Technologies and Systems. 1993.

[WaRK95] R.Y. Wang, M.P. Reddy, H.B. Kon: Towards quality data: An attribute-based approach. Decision Support Systems, **13**(3/4), pp. 349–372, 1995.

[WaSF95] R.Y. Wang, V.C. Storey, C.P. Firth: A framework for analysis of data quality research. IEEE Transactions on Knowledge and Data Engineering, **7**(4), pp. 623–640, August 1995.

[WaSG94] R.Y. Wang, D. M. Strong, L.M. Guarascio: Beyond accuracy: What data quality means to data consumers. Technical report, no. TDQM-94-10, Total Data Quality Management Research Program, MIT Sloan School of Management, Cambridge, MA, 1994.

[WaWa96] Y. Wand, R.Y. Wang: Anchoring data quality dimensions ontological foundations. Communications of the ACM, **39**(11), pp. 86–95, November 1996.

[WaWe90] Y. Wand, R. Weber: An ontological model for an information system. IEEE Transactions on Software Engineering, **16**(11), pp. 1282–1292, November 1990.

[Weld97] Jay-Louise Weldon: Warehouse cornerstones. Byte, pp. 85–88, January 1997.

[WGL*96] J. L. Wiener, H. Gupta, W. J. Labio, Y. Zhuge, H. Garcia-Molina, J. Widom: A system prototype for warehouse view maintenance. In: [MuGu96], pp. 26–33.

[WiCe95] J. Widom, S. Ceri: Active database systems: Triggers and rules for advanced database processing. Morgan Kaufmann, 1995.

[Wido95] J. Widom: Research problems in data warehousing. In: Proceedings of the 4th International Conference on Information and Knowledge Management (CIKM), Baltimore, MD, pp. 25–30. ACM Press, 1995.

[Wied92] G. Wiederhold: Mediators in the architecture of future information systems. IEEE Computer, **25**(3), pp. 38–49, 1992.

[Will97] J. Williams: Tools for traveling data, DBMS and internet systems. DBMS Magazine, **10**(7), p. 69. Also http://www.dbmsmag.com, 1997.

[WizR98] Home Page of WizRule tool available at http://www1.wizsoft.com/index.html, 1998.

[X/Op92] X/Open: Data Management Call Level Interface (CLI). X/Open Snapshot N°S203, Berkshire, UK, September, 1992.

[YaKL97] J. Yang, K. Karlapalem, Q. Li: Algorithms for materialized view design in data warehousing environment. In: [VLDB97], pp. 136–145.

[YaLa87] H.Z. Yang, P.-A. Larson: Query transformation for PSJ queries. In: [VLDB87], pp. 245–254.

[YaLa94] W.P. Yan, P.-A. Larson: Performing group-by before join. In: Proceedings of the 10th International Conference on Data Engineering (ICDE), Houston, TX, pp. 89–100. IEEE Computer Society, 1994.

[YaLa95] W.P. Yan, P.-A. Larson: Eager aggregation and lazy aggregation. In: [VLDB95], pp. 345–357.

[YaWi97] J. Yang, J. Widom: Maintaining temporal views over non-historical information sources for data warehousing. Technical note, http://www-db.stanford.edu/warehousing/publications.html, 1997.

[ZGHW95] Y. Zhuge, H. Garcia-Molina, J. Hammer, J. Widom: View maintenance in a warehousing environment. In: [SIGM95], pp. 316–327.

[ZhGW96] Y. Zhuge, H. Garcia-Molina, J. Wiener: The Strobe algorithms for multi-source warehouse consistency. In: Proceedings of the International Conference on Parallel and Distributed Information Systems, Miami, FL. 1996.

[ZhHK96] G. Zhou, R. Hull, R. King: Generating data integration mediators that use materialization. Journal of Intelligent Information Systems, **6**(2), pp. 199–221, 1996.

[ZHKF95] G. Zhou, R. Hull, R. King, J. Franchitti: Data integration and warehousing using H20. IEEE Data Engineering Bulletin, Special Issue on Materialized Views and Data Warehousing, **18**(2), pp. 29–40, 1995.

[ZHKF95a] G. Zhou, R. Hull, R. King, J. Franchitti: Using object matching and materialization to integrate heterogeneous databases. In: Proceedings of the 3rd International Conference on Cooperative Information Systems (CoopIS), Vienna, Austria, pp. 4–18. 1995.

[ZhWG97] Y. Zhuge, J. Wiener, H. Garcia-Molina: Multiple view consistency for data warehousing. http://www-db.stanford.edu/warehousing/ publications.html, 1997.

Appendix A. ISO Standards Information Quality

The International Organization for Standardization (ISO) was founded in Geneva, Switzerland, in 1946. Its goal is to promote the development of international standards to facilitate the exchange of goods and services worldwide.

The original ISO 9000 [ISO92, ISO97] standards were a series of international standards (ISO 9000, ISO 9001, ISO 9002, ISO 9003, ISO 9004), developed by ISO Technical Committee 176 (TC176) to provide guidance on the selection of an appropriate quality management program (system) for supplier's operations. The series of standards serves the purpose of common terminology, definition, and demonstration of a supplier's capability of controlling its processes. The content of the 1994 edition of the ISO 9000 series is described in the following paragraphs.

ISO 9000-1, Quality Management and Quality Assurance Standards – Guidelines for Selection and Use. This standard explains fundamental quality concepts, defines key terms, and provides guidance on selecting, using, and tailoring series. Furthermore, it helps in the selection and use of the standards in the ISO 9000 family.

ISO 9001-1, Quality Systems – Model for Quality Assurance in Design/ Development, Production, Installation, and Servicing. This is the most comprehensive standard. It addresses all elements including design. The 1994 edition improved the consistency of the terminology and clarified or expanded the meaning of some of the clauses. Several new requirements, such as that for quality planning, were added. The standard contains 20 elements describing the quality parameters, from the receipt of a contract through the design/delivery stage, until the service required after delivery.

ISO 9002, Quality Systems – Model for Quality Assurance in Production and Installation and Servicing. Identical to ISO 9001 except for design requirements. Consequently, it addresses organizations not involved in the design process.

ISO 9002, Quality Systems – Model for Quality Assurance in Final Inspection and Test. This is the least comprehensive standard. It addresses the detection and control of problems during final inspection and testing. Thus, it is not a quality control system. The 1994 edition added additional requirements including: contract review, control of customer supplied product, corrective actions, and internal quality audits.

ISO 9004-1, Quality Management and Quality System Elements – Guidelines. This standard provides guidance in developing and implementing an internal quality system and in determining the extent to which each quality system element is applicable. The guidance in ISO 9004-1 exceeds the requirements contained in ISO 9001, ISO 9002, and ISO 9003. ISO 9004-1 is intended to assist a supplier in improving internal quality management procedures and practices. Yet, it is not intended for use in contractual, regulatory, or registration applications.

Out of them, there is just one, "*ISO/DIS 9000-3 Quality management and quality assurance standards – Part 3: Guidelines for the application of ISO 9001:1994 to the development, supply, installation and maintenance of computer software (Revision of ISO 9000-3:1991)*" specifically intended for use in the computer software industry. Furthermore, there are several standards developed from ISO that are concerned with the achievement of quality in the development and evaluation of software. Yet, these standards are not directly concerned with ISO 9000.

The interested reader can find a lot of other standards developed from ISO and IEEE in the field of software quality. A list of them follows. Note that standards are constantly being added and revised, so this list can quickly become out of date.

IEEE Standards on Information Technology [IEEE 97]

730-1989 IEEE Standard for Software Quality Assurance Plans (ANSI).
1061-1992 IEEE Standard for a Software Quality Metrics Methodology.
730.1-1995 IEEE Standard for Software Quality Assurance Plans. (Revision and redesignation of IEEE Std 938-1986).
1074-1995 IEEE Standard for Developing Software Life Cycle Processes.
1074.1-1995 IEEE Guide for Developing Software Life Cycle Processes.

ISO REFERENCES [ISO 97]

ISO/DIS 9000-3 Quality management and quality assurance standards – Part 3: Guidelines for the application of ISO 9001:1994 to the development, supply, installation and maintenance of computer software.
ISO/IEC 12119:1994 Information technology – Software packages – Quality requirements and testing.
ISO/IEC 9126:1991 Information technology – Software product evaluation – Quality characteristics and guidelines for their use.
ISO/IEC DIS 13236 Information technology – Quality of service – Framework.
ISO/IEC DTR 15504-2 Software Process Assessment – Part 2: A reference model for processes and process capability (normative).
ISO/IEC DTR 15504-3 Software Process Assessment – Part 3: Performing an assessment (normative).

ISO 9000 family

ISO 9000-1: 1994 Quality management and quality assurance standards – Part 1: Guidelines for selection and use.
ISO 9000-2: 1993 Quality management and quality assurance standards – Part 2: Generic guidelines for the application of ISO 9001, ISO 9002, and ISO 9003.
ISO/FDIS 9000-2 Quality management and quality assurance standards – Part 2: Generic guidelines for the application of ISO 9001, ISO 9002, and ISO 9003 (Revision of ISO 9000-2:1993).
ISO 9000-3: 1991 Quality management and quality assurance standards – Part 3: Guidelines for the application of ISO 9001 to the development, supply, and maintenance of software.

ISO/DIS 9000-3 Quality management and quality assurance standards – Part 3: Guidelines for the application of ISO 9001:1994 to the development, supply, installation and maintenance of computer software (Revision of ISO 9000-3: 1991).

ISO 9000-4: 1993 Quality management and quality assurance standards – Part 4: Guide to dependability program management.

ISO 9001: 1994 Quality system-model for quality assurance in design, development, production, installation, and servicing.

ISO 9002: 1994 Quality system-model for quality assurance in production, installation, and servicing.

ISO 9003: 1993 Quality Systems-Model for quality assurance in final inspection and test.

ISO 9004-1: 1994 Quality management and quality system elements – Part 1: Guidelines.

ISO 9004-2: 1991 Quality management and quality system elements – Part 2: Guidelines for services.

ISO 9004-3: 1993 Quality management and quality system elements – Part 3: Guidelines for processed materials.

ISO 9004-4: 1993 Quality management and quality system elements – Part 4: Guidelines for quality improvement.

ISO 10005: 1995 Quality management – Guidelines for quality plans (formerly ISO/DIS 9004-5).

ISO/FDIS 10006 Quality management – Guidelines to quality in project management (Formerly CD 9004-6).

ISO 10007: 1995 Quality management – Guidelines for configuration management.

ISO 10011-1: 1990 Guidelines for auditing quality systems. Part 1: Auditing.

ISO 10011-2: 1991 Guidelines for auditing quality systems. Part 2: Qualification criteria for quality systems auditors.

ISO 10011-3: 1991 Guidelines of auditing quality systems. Part 3: Management of audit programs.

ISO 10012-1: 1992 Quality assurance requirements for measuring equipment – Part 1: Metrological confirmation system for measuring equipment.

ISO 10013 Guidelines for developing quality manuals.

ISO/TR 13425 Guidelines for the selection of statistical methods in standardization and specification.

ISO 8402: 1994 Quality management and quality assurance – Vocabulary.

Appendix B. Glossary

B.1 Data Warehouse Systems

Agent: A program that controls components and processes in the data warehouse.

Centralized data warehouse architecture: An architecture where all the data are stored in a single database system, directly accessible to all client applications.

CWM (Conceptual Warehouse Model): A metamodel proposed as a standard by OMG to support design and operation of ROLAP- and MOLAP-based data warehouse solutions.

Data cleaning: Process of cleaning the source data from inconsistent values during the loading process.

Data mart: A data store which stores a subset or aggregation of the data of the data warehouse. A data mart can be seen as a small local data warehouse.

Data source: A system from which data are collected, in order to be integrated into the data warehouse.

Data warehouse component: A building block of the data warehouse, e.g., database systems, agents, client applications.

Dimension table: Table of star or snowflake schema which contains the possible values of one dimension.

Fact table: The central table in a star or snowflake schema which contains the measures of interest.

Federated data warehouse architecture: An architecture where the data are logically consolidated but physically distributed among various data marts.

Integrator: A program that integrates the data that are collected by wrappers from several data sources.

Loader: In the context of data warehouses, a synonym for wrapper.

Mediator: A program to integrate several data sources into the data warehouse.

Meta database: A repository for metadata of the data warehouse.

Metadata: Any description of the data warehouse components (e.g., their schema) and the relationship between data warehouse components.

Metamodel: A framework for representing common properties of conceptual modeling languages for metadata.

MOLAP: Multidimensional OLAP, that means building OLAP applications on top of a multidimensional database.

Multidimensional database system: A database system that provides a multidimensional view – and possibly storage – for its data.

OIM (Open Information Model): A metamodel by the Metadata Coalition proposed as a standard for lifecycle-wide information modeling support.

OLAP (Online Analytical Processing): Interactive analysis of data that has been transformed from the raw (operational) data into understandable enterprise-wide data.

OLTP (Online Transaction Processing): Transaction-oriented work with operational systems.

Primary data warehouse: A database in the data warehouse where *all* source data are integrated and collected.

Repository: Database for the metadata of the data warehouse.

ROLAP: Relational OLAP, that means building OLAP applications on top of a relational database.

Schema: The representation of data in a data warehouse component.

Secondary data warehouse: A database dedicated to specific client applications. The data from the primary data warehouse may be extracted and aggregated in the secondary data warehouse.

Snowflake schema: Refinement of star schema with less redundancy in relations.

Star schema: Schema to represent multidimensional data in a denormalized relational database.

Tiered data warehouse architecture: An architecture where the data are stored on different tiers. Each tier is a summarization of the previous tier.

Wrapper: A program that reads data from a data source and stores them in the data warehouse. Data transformation and data cleaning are possible during this process.

B.2 Data Quality

Aliases: A description of the alias names for several fields in the sources and the data warehouse.

Benchmarking: A systematic method by which organizations can measure themselves against the best industry practices.

Data accuracy: A description of the accuracy of the data entry process which happened at the sources.

Data completeness: A description of the percentage of the information stored in the sources and the data warehouse with respect to the information in the real world.

Data consistency: A description of the consistency of the information which is stored in the sources and the data warehouse.

Data credibility: A description of the credibility of the source that provided the information.

Data currency: A description of when the information was entered in the sources (data warehouse). In a temporal database, currency should be represented by transaction time.

Data non-volatility: A description of the time period for which the information is valid. In a temporal database, currency should be represented by valid time.

Data origin: A description of the way a data warehouse relation (view) is calculated from the data sources.

Data quality assurance: All the planed and systematic actions necessary to provide adequate confidence that a data product will satisfy a given set of quality requirements.

Data quality control: A set of operational techniques and activities which are used to attain the quality required for a data product.

Data quality management: The management function that determines and implements the data quality policy.

Data quality policy: The overall intention and direction of an organization with respect to issues concerning the quality of data products.

Data quality system: The integration of the organizational structure, responsibilities, procedures, processes and resources for implementing data quality management.

Data semantics: A description of the semantics of a relation and of each one of its attributes.

Data syntax: A description of the type of each attribute of a relation, the primary and foreign keys, the triggers and the stored procedures, etc. The syntax dimension is also known as data dictionary.

Data usage: The profile of the use of the information stored in the data warehouse.

GQM (Goal-Question-Metric approach): A software engineering methodology, originally developed for software quality management. In GQM, the high-level user requirements are modeled as goals. Quality metrics are values which express some measured property of the object. The relationship between goals and metrics is established through quality questions.

Minimality: A description of the degree up to which undesired redundancy is avoided during the data extraction and source integration process.

Process completeness: A description of the completeness of the process with regard to the information which is unintentionally ignored.

Process consistency: A description of the consistency of the process with regard to the uniqueness and noncontradiction of the information in a data warehouse.

Process description: A description of the algorithm (data flow and transformations) and the data sources which are used for the calculation of the data warehouse views.

QFD (Quality Function Deployment): A team-based management tool, used to map customer requirements to specific technical solutions. This philosophy is based on the idea that the customer expectations should drive the development process of a product.

Quality dimensions: A set of attributes of the data or the processes of a warehouse, by which quality is described, in a high-level, user-oriented manner.

Quality goal (GQM): A high level, conceptual intention of a user, regarding the quality of the system he/she deals with, defined for an *object*, for a variety of *reasons*, with respect to various *models of quality*, from *various points of view*, relative to a *particular environment*.

Quality metrics (GQM): a set of data, associated with every question in order to answer it in a quantitative way.

Quality question (GQM): A set of questions is used to characterize the way the assessment/achievement of a specific quality goal is going to be performed, based on some characterizing model. Questions try to characterize the object of measurement (product, process, resource) with respect to a selected quality issue and to determine its quality from the selected viewpoint.

Quality: The fraction of the performance of a product or service with regard to the expectation the user has towards this performance. Alternatively, one can think of quality as the loss imparted to society from the time a product is shipped. The total loss of society can be viewed as the sum of the producer's loss and the customer's loss.

Relevancy to data warehouse: A description of the relevancy of each attribute (relation) of the source data to the data warehouse.

Replication rules: A description of the replication rules existing in the warehouse.

System availability: A description of the percentage of time the source or data warehouse is available.

Total quality management: A philosophy for the improvement of an organization in order to achieve excellence.

Transaction availability: A description of the percentage of time each record is available, due to the absence of update operations in the sources or the data warehouse.

Update frequency: A description of the frequency of the update process for the batch and periodic modes.

Update mode: A description of the update policy for each data warehouse view (e.g., batch, periodic, on demand, immediate).

User privileges: A description of the read/write privileges of each user for a certain relation of the source (data warehouse) data.

Validation: A test on the correctness of each process.

Version control: A description of the metadata evolution of the data warehouse.

B.3 Source Integration

Conceptual data warehouse schema: A conceptual description of the data stored in the data warehouse.

Data integration: The process of comparing a collection of source data sets, and producing a data set representing a reconciled view of the input sources, both at the intentional (schema) and the extensional (data) level.

Extensional wrapper: A program that extracts data according to the specification given by a corresponding intentional wrapper. The data which are the output of extensional wrappers have a common, prespecified format.

Intentional wrapper: A mapping that specifies how information represented in a source map to the concepts in the conceptual data warehouse schema, and how data are to be extracted from the sources.

Interschema assertion: specification of a correspondence between a certain data set in one source and another data set in another source.

Mediator: a module that processes the output of extensional wrappers, by cleaning, reconciling, merging data, and storing the resulting data in the data warehouse.

Schema integration: the process of comparing source data schemata, and producing as output a new data schema (target schema) representing the reconciled intentional representation of the input schemata.

Source description: a description of the source data which are of interest to the data warehouse.

B.4 Multidimensional Aggregation

Aggregation – consolidate – roll-up: The querying for summarized data. Aggregation involves computing the data relationships (according to the attribute hierarchy within dimensions or to cross-dimensional formulas) for one or more dimensions. For example, sales offices can be rolled-up to districts and districts rolled-up to regions; the user may be interested in total sales or percent-to-total.

Aggregation function: Functions – like, e.g., sum, min, max, average, count – computing an aggregated measure from a set of basic values belonging to the same dimension.

Concrete domain: A domain (such as the integers, reals, enumeration types, etc.) on which aggregation functions and predicates are defined, together with the corresponding functions and predicates themselves. It is clearly distinguished from the abstract domain modeled on the logical level.

Dimension: A dimension is a structural attribute acting as an index for identifying measures within a multidimensional data model. A dimension is basically a domain, which may be possibly partitioned into a hierarchy of levels. For example, in the context of selling goods, possible dimensions are product, time, and geography; chosen dimension levels may be Product category, Month, and District.

Dimensional modeling: A technique within ROLAP to organize information into two types of data structures: measures, or numerical data (for example, sales and gross margins), which are stored in "fact" tables; and dimensions (for example, fiscal quarter, account and product category), which are stored in satellite tables and are joined to the fact table.

Level: A partitioning of a dimension defines the various levels for that dimension. For instance, a spatial dimension might have a hierarchy with levels such as country, region, city, or office. A set of levels of different dimensions defines a hypercube for a measure depending on those dimensions. For example, "sales volume" can be a measure depending on the levels Product category (product dimension), Month (time dimension), and District (geographical dimension).

Measure – variables – metrics: A measure is a point into the multidimensional space. A measure is identified if for each dimension a single value is selected. For example, a "sales volume" measure is identified by giving a specific product, a specific sale time, and a specific sale location.

Multidimensional data model: The way multidimensional information is abstractly represented. A multidimensional data model is a *n*-dimensional array, i.e., a hypercube. A cell in this *n*-dimensional space is a measure seen as depending on the n dimensions.

Multidimensional information: The information is "multidimensional," if the data can be seen as depending on several independent variables. This means for the user that it can be visualized in grids: information is typically displayed in cross-tabs, and tools provide the ability to pivot the axes of the cross-tabulation. For example, "sales volume" is a multidimensional information if viewed as a function of a set of dimensions, e.g., product, time, and geography.

Navigation – slicing-and-dicing: The processes employed by users to explore and query multidimensional information within a hypercube interactively.

Pivot – rotate: To change the dimensional orientation of the cube, for analyzing the data using a particular dimension level as independent variable. For example, rotating may consist of swapping the rows and columns, moving one of the row dimensions into the column dimension, or swapping an off-spreadsheet dimension with one of the dimensions in the page display (either to become one of the new rows or columns), etc. A specific example of the first case would be taking a report that has Time across (the columns) and Products down (the rows) and rotating it into a report that has Product across and Time down. An example of the second case would be to change a report which has Measures and Products down and Time across into a report with Measures down and Time over Products across. An example of the third case would be taking a report that has Time across and Product down and changing it into a report that has Time across and Geography down.

Query answering: The retrieval of all those instances in a given database which satisfy certain properties given by the query.

Query containment: Containment of queries is the problem whether one query is more general than another one, that is, whether each answer of the latter query is always also an answer of the more general one.

Query refinement: Refinement of queries is the problem checking whether a query involving aggregation and a (materialized) view can be computed using (the aggregations contained in) the view itself. This depends on whether the aggregations contained in the view are still fine-grained enough to compute the aggregations required by the query. For example, suppose a user asks the system to compute a query about the total profit of all product groups for each year and each region. If a (materialized) view exists which contains the profit for the product groups Food and Nonfood for all quarters for all regions, then the total profit can be computed by simply summing up those partial profits for each year.

Query satisfiability: Satisfiability of queries is the problem whether there exists a world such that each query of a given set of query has a nonempty answer.

Roll down – drill down: The navigation among levels of data ranging from higher level summary (up) to lower level summary or detailed data (down). The drilling paths may be defined by the hierarchies within dimensions or other relationships that may be dynamic within or between dimensions. An example query is for a particular product category, find detailed sales data for each office by date.

Scoping: Restricting the view of database objects to a specified subset. Further operations, such as update or retrieve, will affect only the cells in the specified subset. For example, scoping allows users to retrieve or update only the sales data values for the first quarter in the east region, if that is the only data they wish to receive.

Screening – selection – filtering: A criterion is evaluated against the data or members of a dimension in order to restrict the set of data retrieved. Examples of selections include the top ten salespersons by revenue, data from the east region only and all products with margins greater than 20%.

Slicing: Selecting all the data satisfying a condition along a particular dimension while navigating. A slice is a subset of a multidimensional array corresponding to a single value for one or more members of the dimensions not in the subset. For example, if the member Actuals is selected from the Scenario dimension, then the subcube of all the remaining dimensions is the slice that is specified. The data omitted from this slice would be any data associated with the nonselected members of the Scenario dimension, for example Budget, Variance, Forecast, etc. From an end-user perspective, the term slice most often refers to a two-dimensional page selected from the cube.

B.5 Query Optimization

Aggregate navigator: In data warehouse architectures with precomputed aggregate views, the aggregate navigator mediates between the query tools and the DBMS. The aggregate navigator contains information about which aggregate views are materialized and where they are stored. It tests whether incoming queries can be rewritten using the existing views. Among the possible rewritings, the aggregate navigator chooses one with minimal computation cost. Aggregate navigators perform a limited test for view usability.

Bitmap index: A bitmap index for an attribute a of a relation r represents the location of records by lists of bits. More precisely, for each value v of attribute a, there is a list of bits having 1 at position i if the i-th record satisfies $ri.a = v$ and having 0 elsewhere. Bitmaps are well suited for data warehouses, since many attributes on dimensions have value sets of small cardinality. They can be generated efficiently when data are loaded. The most important benefit of bitmap indexes is that constraints in a query that are Boolean combinations of simple constraints on attribute values can be evaluated with bitmap indexes alone, by intersections, unions, or complements of the lists, without accessing the relation.

Business report: End users should be allowed to view the contents of a data warehouse in terms of their business perspective. A convenient way for end users to specify their information needs is to define a business report. The values in the table are computed by aggregating base data at different levels. Business reports typically contain different types of comparisons between aggregates. Query tools provide end users with high level primitives to define business reports.

Canned queries vs. Ad-hoc queries: In an OLTP system, queries are the interface between the database and application programs. Therefore, queries are designed by programmers and are rigidly built into the system. Usually, they allow some variation in that some parameters in the query can vary. Queries of this type are called preformulated or canned queries. In data warehouses, loading operations and managerial interfaces are defined by canned queries.

Many queries, however, are created through query tools by end users that explore the content of the warehouse. Canned and ad-hoc queries demand different techniques for optimization. While the former can be optimized at compile or even design time, for the latter the time available for optimization should be within fractions of the run time of the query.

Indexing: Indexes are auxiliary structures that allow a fast access to records in a relation or to certain values in those records. In OLTP databases, the usage of indexes is restricted because an index on a relation has to be modified after each update of the relation. Data warehouses, however, with their read-only environment, offer much more options to use indexing techniques. A value list index for an attribute maintains for each value of the attribute a list of pointers to the records with that particular value of the attribute. For attributes with few distinct values, bitmap-indexes are more space efficient and yield considerable speed-ups. A projection index for an attribute provides fast access to the values of the attribute without accessing the entire record. By using both kinds of index simultaneously, answers for certain queries over a relation can be computed without accessing the relation itself.

Join index: When querying a star schema, indexes on a single table are of limited use, because queries join the fact table with the dimension tables. A star schema can be seen as a normalized representation of a single virtual relation – the one that results from joining the fact table with all the dimension tables. A join index then is an index on that virtual table. It associates the value v of an attribute a on the dimension table Rd to those records in the fact table Rf that join with a record rd in Rd that satisfies $rd.a = v$. With join indexes, join queries can be evaluated without computing the joins. Join indexes can be of any of the types discussed before: value-list, projection, and bitmap.

Query tool: Query tools provide access to the data warehouse for users not familiar with formal query languages. They offer graphical interfaces that allow one to edit complex business reports by point and click operations. If the query tool interfaces a relational implementation of a data warehouse, the business report is broken down into several SQL queries that are shipped to the data warehouse server. The query tool receives the individual answer sets and stitches them together to an answer of the business report. Ideally, query tools realize a bottleneck architecture, i.e., they ship small queries to the server and receive only small answer sets.

View usability: When a new query is to be optimized in an environment where answers to other queries (= views) are materialized, the basic question is whether any of the precomputed query results can be used for answering the new query. When views and queries are conjunctive queries, the view usability problem is NP-complete. Techniques to solve the view usability problem for conjunctive queries have been extended to SQL queries with aggregates. However, there is no comprehensive set of results for the view usability problem in data warehousing.

Index